MEL BARTHOLOMEW
FOUNDATION

GROWING PERFECT

veget

ables

a visual guide
to raising and
harvesting
prime garden
produce

COOL
SPRINGS
PRESS

MINNEAPOLIS, MINNESOTA

© 2017 Quarto Publishing Group USA Inc.

First published in 2017 by Cool Springs Press, an imprint of The Quarto Group, 401 Second Avenue North, Suite 310, Minneapolis, MN 55401 USA. Telephone: (612) 344-8100 Fax: (612) 344-8692

quartoknows.com
Visit our blogs at quartoknows.com

10 9 8 7 6 5 4 3 2 1

ISBN: 978-1-59186-683-1

Library of Congress Cataloging-in-Publication Data

Names: Cool Springs Press, issuing body.
Title: Square foot gardening : growing perfect vegetables : a visual guide to raising and harvesting prime garden produce.
Description: Minneapolis, MN : Cool Springs Press, 2017.
Identifiers: LCCN 2016045691 | ISBN 9781591866831 (pb)
Subjects: LCSH: Food crops–Ripening. | Horticultural crops–Ripening. | Square foot gardening.
Classification: LCC SB175 .S68 2017 | DDC 635–dc23
LC record available at https://lccn.loc.gov/2016045691

Acquiring Editor: Todd R. Berger
Project Manager: Alyssa Bluhm
Art Director and Cover Design: Cindy Samargia Laun
Interior Book Design: Ashley Prine, Tandem Books

Printed in China

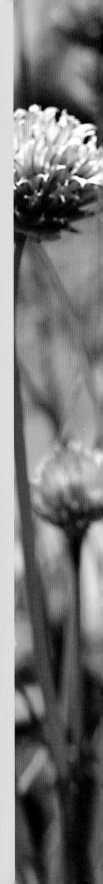

CONTENTS

INTRODUCTION

WHEN MEL BARTHOLOMEW FIRST THOUGHT UP THE SQUARE FOOT
Gardening system in 1975, he was frustrated. He had seen a lot of ripe produce
go to waste in his local community garden simply because of inefficiency
and gardener burnout. Mel was a math-minded engineer to the core, and he
believed gardening could be made more efficient, with success that could be

put into numbers. He wanted his garden to
produce more ripe edibles per square foot
than any other. He thought it should require
measurably less effort and inputs than any
other garden. The numbers had to make
sense. He was a man on a mission.

Of course, as a gardener, he also
wanted a nice-looking, self-contained bed
that didn't need any of that awful back-
breaking weeding. He considered himself an
environmentalist, and wanted his gardening
method to use 90 percent less water than
conventional row gardens did, and it had to
do fine without pesticides or fertilizers. Like

all gardeners, Mel preferred the fun and flavor of gardening to the repetitive
hard work. He wanted his garden method to capture the enjoyment, not the
drudgery, of gardening.

∧ Square Foot Gardening founder Mel Bartholomew.

< Square Foot Gardening limits harvests to what you can realistically use and makes it easier to
determine ripeness.

Square Foot Gardening turned out to be all that and more.

But even if you grow the perfect Square Foot Garden for you and your family and take advantage of all that mathematical thinking behind the method, you're still missing out if you're not picking your produce at exactly the right time. (Or, as Mel liked to say, "The *ripe* moment!")

The fact was, Mel *could* teach people how to build a Square Foot Garden raised bed, he *could* show them how to make the Mel's Mix soil they'd need, and he *could* even demonstrate the proper way to plant inside a square. But he simply *couldn't* hang around to show gardeners exactly when their produce was ready to be harvested. That information has been a missing link in the chain that leads from a Square Foot Garden to kitchen and table. Until now, that is.

This book takes the guesswork out of picking—and selecting already-picked—perfectly ripe produce. The information applies whether you're growing a Square Foot Garden, another type of raised-bed garden, a container garden, a row garden, or just buying most of your produce from a market.

Let's face it, we've all dealt with the disappointment of produce that turns out to be too unripe or too overripe to eat or cook. Almost everyone at one time or another has gotten an avocado home only to find out it was so far gone that it had turned to brown mush inside. Ripening can be a bit of a mystery.

COOKED OR NOT?

Although raw-food diet proponents promote eating vegetables in their natural state, there's a good reason why you should consume both raw and cooked vegetables. Eaten raw, vegetables such as carrots provide excellent vitamins and nutrients, and the fiber goes through your digestive system without breaking down. But cooking carrots, broccoli, and most other vegetables destroys cell walls, often making crucial nutrients and compounds available to the body when they otherwise wouldn't be. This is true of the essential beta-carotene in carrots and some of the cancer-fighting compounds in broccoli and other cruciferous vegetables. Cooking also can enhance sweetness and flavor in ripe vegetables.

That's why the first chapter in this book offers a brief, basic, and understandable explanation of what ripening is (and isn't). That explanation will give you a strong grasp of what exactly is going on inside fruits and vegetables—what it is that determines when they should come off the vine or out of the ground.

But the centerpiece to this book is Chapter 2: "The Ripeness Listings." We've broken this list into fruits and vegetables commonly grown in an SFG, those that aren't grown "in the box" because they grow on plants too large for a square foot or because they ripen on a tree, and a third section for fruits and vegetables you'll most likely only find in the produce aisle at your local grocery store. Each individual listing, however, provides all the insight you'll need to determine if a fruit or vegetable is ripe, no matter where you might find it. The listings also cover how to actually harvest the fruit or vegetable if that's appropriate, and tips on making the most of the ripe produce you bring into the kitchen.

Of course, once you get to the point of harvest, there are a lot of varied ways to take advantage of ripe produce and even extend the life of already ripe produce. That's why the book wraps up with a short chapter on manipulating ripeness, including ways to make your fruits and vegetables last as long as possible, or methods for using them when ripe and then storing the finished product.

RIPENING THE SFG WAY

Square Foot Gardening is a great way to detect ripeness in your fruits and vegetables, because the garden is contained in such controlled, small spaces that it doesn't take a lot of hunting to see the state of your edibles.

Regular interaction with your plants and basic garden maintenance is your chance to monitor how your crops are ripening, and how to determine when to harvest at just the right moment.

Based on growing one type of plant per square in a 4 × 4-foot box divided into 16 squares, Square Foot Gardening swept not only America, but the world when it was introduced. It started simply enough, with Mel trying to make a community garden more successful and easier to maintain. The idea grew in popularity until Mel had more people interested in this new way of gardening than he could reasonably teach. That led to him write *Square Foot Gardening: A New Way to Garden in Less Space With Less Work* in 1981. A PBS series followed, which in turn was followed by more books, including the most recent, *Square Foot Gardening High-Value Veggies: Homegrown Produce Ranked by Value.*

SFG boxes are filled with a soil blend made up of one-third peat, one-third vermiculite, and one-third blended compost. This ensures that the soil is weed-free from the start and that square foot gardeners won't need to weed. The mix is also great at retaining moisture, so SFGs are incredibly water efficient.

The SFG box is intentionally sized to give gardeners the ability to reach the center from all sides. This makes harvesting ripe fruit and vegetables simple; they're always within reach.

Of course, Mel also taught his students that a key concept in Square Foot Gardening is to calculate and grow only what you and your family really consume. Mel was a fun-loving, upbeat guy, but nothing could rile him more than seeing an oversized garden with valuable produce rotting on the vine because the gardener grew too much.

That's why it's so important to know as much as possible about ripening. Some crops ripen all at once, while others deliver a steady stream of edibles over the season. So many gardeners wind up with a kitchen full of zucchini or apples and no idea what to do with the extra. A little bit of planning goes a long way toward ensuring none of your harvest goes to waste. The squares let you do that easily, as described in the book *All New Square Foot Gardening*, but you should keep that in mind even if you're row gardening, container gardening, or using your own raised-bed system.

Is it ripe? You bet it is.

The other way to make the most of seasonal ripening is succession planting—planting crops that mature in spring, followed by crops that mature in summer, and so on. That, too, was a big part of Mel's Square Foot Gardening and one you'll find described in Chapter 3.

Whatever strategies you use, making the most of ripe produce is essential for enjoying gardening as you should, and for getting the most out of your gardening and produce budget. Ultimately, you'll save money, time, effort, and frustration if you can answer one simple question: is it ripe?

This book gives you the answer.

> A Square Foot Garden looks most natural when the plants are just maturing to the stage of ripeness.

CHAPTER 1

WHAT IS RIPE?

IF YOU'RE GOING TO GO TO ALL THE EFFORT IT TAKES TO PLANT, TEND, and manage a garden—and certainly if you're spending your hard-earned money on fresh produce—you deserve to pick the most perfectly ripe fruits and vegetables possible. Knowing how produce ripens is key not only to harvesting or buying the best produce, but also to controlling the ripening process for fruits and vegetables that ripen off the vine, and others that can be affected by them.

> **Few feelings can match the satisfaction of harvesting ripe produce fresh from your garden.**

Mel Bartholomew always said, "Start at the beginning." So the first step in understanding ripening is the definition: Ripe is the stage of growth or maturation in which any fruit or vegetable is at its ideal point to be eaten. Makes sense, doesn't it? If the goal of edibles is to be eaten, ripeness is the peak of flavor and texture, and the point at which nutrients are most concentrated.

That's a good starting point, but keep in mind that some plants, such as greens and many root vegetables, can be picked when immature and still deliver plenty of value. They may not be totally "ripe" but they are still desirable. So the definition doesn't cover the whole issue of ripe. Things are just a little bit more complicated than the definition makes them out to be.

GARDEN WISDOM: COMPANION PLANTING

At first glance, companion plants may not seem to have anything to do with ripeness, but they're more relevant than you might think. Planting natural companions to garden fruits and vegetables helps protect those crops against insect and disease problems. But it's just as important to avoid planting your edibles with certain plants known in gardening circles as "enemy plants." Enemies can attract pests that will attack the companion plant and, more often, they compete with the plant for nutrients, leading to more feeble harvests than normal.

- **Collard green** enemies are celery, potatoes, and yams.
- **Beans and peas** shouldn't be grown next to garlic, onions, or shallots, because the ripe beans could be stunted.
- **Corn and tomatoes** should be separated, because both are attacked by the same worm.
- **Cucumbers** should be grown away from sage, because the herb harms cucumber plants.
- **Strawberries** should be planted far from cabbage, because they compete for nutrients.

The idea of ripe applies to pretty much any crop you can grow or buy. But how and when a fruit or vegetable gets ripe is another matter entirely. Most fruits and all vegetables have to ripen on the plant, because their growth stops the moment they're picked (these are called *non-climacteric*). After that point, they get no riper, even though they still age, deteriorating in quality as they do. Other fruits continue to ripen naturally even after they've been removed from the plant (these are called *climacteric*). That's why you can buy unripe bananas and they'll peak in flavor and texture a few days later, while a watermelon will never get any riper than the day you pick it (so you better know it's ripe when you harvest or buy it!).

The difference between climacteric and non-climacteric produce is ethylene gas . . . or, actually, the amount of ethylene gas. This gas is produced by virtually all plants to one degree another. But in climacteric fruits, the amount of gas is high and acts as a hormone that stimulates ripening. In other words, climacteric fruits have

> Climacteric or not? Knowing which of these ripens itself will affect how you store it, and whether you can buy the fruit unripe or not.

the mechanism to ripen themselves. That's pretty handy because it means they can be picked unripe and transported to market.

CLIMACTERIC FRUIT	NON-CLIMACTERIC FRUIT
Apple	Asparagus
Apricot	Blackberry
Avocado	Blueberry
Banana	Bell pepper
Fig	Broccoli
Honeydew melon	Cherry
Kiwi	Cucumber
Mango	Hot pepper
Nectarine	Grapes
Papaya	Grapefruit
Peach	Eggplant
Plum	Kale
Tomato	Lemon
	Lettuce
	Lime
	Orange
	Pineapple
	Pomegranate
	Pumpkin
	Raspberry
	Spinach
	Squash
	Strawberry
	Watermelon
	Zucchini

Non-climacteric fruits and vegetables don't ripen in the presence of ethylene gas, but the gas does speed up aging and deterioration of these crops. That's why storing different fruits and vegetables together can be tricky. If you store the wrong combination of fruits and vegetables, some will spoil more quickly than you anticipated.

Ethylene gas is the reason these bananas can make it thousands of miles to your local market without rotting during the trip.

THE MAGIC OF ETHYLENE

Just as hormones control aging in humans, ethylene gas can control ripening in climacteric fruits. But how does it do that? Ethylene breaks down the cell walls of the fruit so that the internal compounds mingle, making the fruit juicier and creating a complex flavor. Of course, once the cell walls are broken down, the fruit loses its structure and becomes softer. If the fruit is not eaten, it will eventually just rot or turn to mush.

Climacteric fruits usually should not be stored enclosed with non-climacteric fruits. Simple enough, right? Put a cut banana in the vegetable drawer with lettuce greens, and the greens will age and deteriorate much faster than they otherwise would. You'll wind up with limp, tasteless salad greens in a day and wonder how that happened.

Of course, the opposite is true too. You can use the principle to serve your own needs. Corporate growers and fruit distributors use ethylene gas to artificially speed up ripening in the produce they're shipping. Bananas come off the boat from Ecuador completely green, and a distributor puts them in a room filled with ethylene gas. The bananas are ready for sale, mostly ripe, in a day or two.

You can actually use this effect yourself to ripen produce that, for one reason or another, you bought or picked unripe. (In fact, you'll find covered "fruit ripening" bowls at retail for just this purpose.) It might have been the end of the season, or maybe they were the only tomatoes left on the produce aisle.

You can speed up the ripening of a pear, for example, by putting it into a paper bag with an apple. (This method also works by submerging the fruit itself in dry rice or covering it in a soft, clean dishtowel.) The apple will emit a wealth of ethylene, which will increase the rate of ripening in the pear. Whatever the case, knowing which fruits and vegetables ripen themselves can be mighty handy. Here's a list of fruit that can be artificially ripened in this way.

CAN BE BAG-RIPENED	CANNOT BE BAG-RIPENED
Avocado	Apple
Banana	Blackberry
Cantaloupe	Blueberry
Honeydew melon	Cherry
Mango	Cranberry
Nectarine	Gooseberry
Papaya	Grapes
Peach	Lemon
Pear	Lime
Persimmon	Orange
Tomato	Pomegranate
	Raspberry
	Strawberry
	Tangerine
	Watermelon

You'll find more information about using climacteric produce to control ripening in the individual listings throughout Chapter 2, and Chapter 3 covers ways to control ripeness and extend the life of the produce you grow or buy.

Vegetables are all non-climacteric and ripen more slowly and in a more controlled fashion than fruit does. That often translates to a longer shelf life (well, longer life in the refrigerator

THE IODINE TEST

Want to see ripe in action? (Or want to show your children a little basic food chemistry?) Try this experiment. Iodine can be used to determine ripening or rotting in fruit because it reacts differently to starch and sugar. Cut an apple in half and put a drop of iodine on the exposed flesh. If it turns black, the fruit hasn't even started to ripen. If it turns blue or black after a couple seconds, ripening is in progress, but not complete. If it turns yellow or orange, ripening is finished, and the fruit is on its way to rotting.

crisper bin). Because of their makeup, vegetables won't release their true flavors, smells (if any), or nutrients until their cells are crushed—either through cutting, chewing, or cooking. Ripening heightens these effects. Don't believe it? Cut open a ripe yellow onion and slice open a scallion of the same cultivar. Take a deep whiff. The difference between the intensity of reaction in the ripe (onion bulb) and unripe (scallion) versions should be painfully clear.

SEASONAL & CONDITIONAL RIPENESS

There's ripe and then there's *ripe*. When and how certain fruits and vegetables ripen can radically affect how they taste. Seasonal factors can play a large part in how a fruit or vegetable ripens, and the same crop may be noticeably different picked ripe in two different seasons.

For instance, many cool-weather crops can be grown in spring and fall, but the fall versions, kissed with frost, will be much sweeter because the colder temperatures trigger the plant to convert starch to sugar. That's because sugar essentially functions as antifreeze inside plant cells. Keep in mind a light frost is one thing. A heavy frost or hard freeze will damage many plants, unless you've applied several inches of mulch or otherwise protected the plants. Here are some examples of crops you may want to plant in autumn:

SWEETER AFTER FROST
Beets
Brussels sprouts
Cabbage
Carrots
Collard greens
Kale
Leeks
Parsnips
Rutabagas
Spinach
Swiss chard
Turnips

Seasons aren't the only factors that make for different notions of ripeness in the same edible. Ripe is relative, regardless of how straightforward the scientific definition may be. Although there is a perfect moment of ripeness for a tomato—a point at which it is exactly between rock-hard and mushy, and the color is a brilliant red with a rich, full flavor—other plants are not so definitive.

Spinach picked young can be delectable because the "baby" spinach leaves are tender, fresh, and lightly flavored. If you want a few simple greens to add to a mixed salad, baby

Many sturdy garden greens become noticeably sweeter after a light frost, but a hard freeze will kill them.

spinach may be just the thing you're after—even though those leaves couldn't be considered scientifically "ripe." Mature spinach has its place in dishes cooked and fresh, and you'll certainly realize more of it if you let it mature, but it's a perfect example of a plant that may have many versions of ripe.

There is an intangible in getting the ripest produce from your garden: diligence. Nothing substitutes a daily check of your plants because so many variables can affect ripeness. A sudden period of hot weather, a variation in the amount of water the plant gets, and even excessive weeds can slow down, speed up, or otherwise affect ripening. This is why the "days to maturity" listing on a seed packet has to be considered a ballpark estimate. There's no way a seed manufacturer can accurately predict exactly when fruits or vegetables will ripen in your garden.

It's not just that your local climate may vary significantly for the norm in your zone. There are just too many variables that can affect plant ripening. Disease and pests are at the top of the list. Some diseases in plants can stop fruit from ripening. Other diseases and conditions can

When it comes to berries, cherries, and tree fruits, birds can't get enough. If you want to protect your share of the harvest, netting is a simple, inexpensive option.

cause fruit or vegetables to ripen imperfectly—not what you want see after putting a lot of effort and resources into your garden.

Pests are part of the picture too. Birds and wildlife like ripe produce every bit as much as humans do. They can be good indicators of when a certain fruit or vegetable is perfectly ripe. However, raccoons can also strip a grapevine, gophers can steal carrots, and birds can decimate a cherry tree amazingly fast. Unless you're okay with sharing your harvest, keeping any eye out for garden thieves is all part of getting your crops to that deliciously ripe perfection.

Optimizing ripening starts with taking steps to ensure plant health. Know how much water your specific plants need and when they need it (some plants, such as watermelon, ripen best when you discontinue watering at the end of the ripening process).

Ultimately, that is why there is simply no substitute to knowing your garden intimately when it comes to harvesting perfectly ripe produce. Mel Bartholomew was a big proponent of gardeners spending time with their plants every day, even when there wasn't anything that needed to be done in the garden.

There are a lot of upsides to this type of attention. You catch disease and pest damage early, giving yourself a chance to address it. You'll notice overly dry soil (or overly wet) and can adjust your watering. You'll gain a sense of how things are progressing and, in becoming more familiar with the unripe state of fruit and vegetables, you'll be keenly aware of that moment when they mature to fully ripe. In short, you'll give yourself a leg up in harvesting ripe crops at the perfect moment.

In the end, it's all about never having to ask, "Is it ripe?" because you'll already know the answer.

> Picking ripe fruit and vegetables is the most rewarding part of gardening for children and adults alike.

CHAPTER 2

THE RIPENESS LISTINGS

THE INDIVIDUAL LISTINGS THAT FOLLOW ARE BROKEN DOWN INTO THREE groups: first, Square Foot Gardening crops are those that are regularly grown in the 16-square-foot SFG "box"; second, out-of-the-box edibles are those that must be grown outside the confines of the SFG box because the plant grows too large, sprawls too much, or is simply not appropriate for the SFG culture; and third, produce you'll only find in the market and in a very limited range of geographic locations in the United States.

> **The ripeness you're looking for often differs between picking something fresh and picking it up at the market . . .**

Regardless of this breakdown, we've included information for harvesting or buying the fruit or vegetable. The ripeness you're looking for often differs between picking something fresh and picking it up at the market; we've tried to cover all bases in these listings.

The listings are meant to cover all common fruits and vegetables. We've excluded those that are simply too rare to commonly find within the United States. That means you'll find just about every crop you can grow or buy in this country.

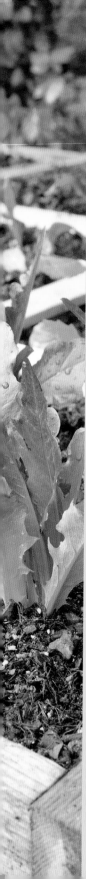

INSIDE THE BOX
Square Foot Gardening Crops

ASPARAGUS

ASPARAGUS CAN BE A TRICKY VEGETABLE to harvest correctly. There is a definite harvest starting and stopping point, and you need to harvest your asparagus regularly to ensure it doesn't grow tough and inedible.

Asparagus takes two years to establish. First-year growth should be harvested lightly, if at all, to ensure a robust harvest in the second year. The first spears begin appearing as soon as the soil warms to 50°F. Harvest when they are between 7 and 9 inches long (measuring from the soil). At the beginning of the season the spears will be thin. As the season progresses, they grow thicker. Use a sharp serrated knife to cut the spears at an angle as close as possible to the soil, or bend and snap the spear off as close to the soil as possible. Harvest in the morning, because the spears lose their freshness quickly in heat of midday.

Once the air temperature warms, harvest every day to avoid individual spears growing

RIPE. Early season asparagus is thin when mature; late-season spears will be thicker. Perfectly ripe asparagus is 7 to 9" tall.

too large—which translates to tough, fibrous spears. Don't eat spears that have begun to open; harvest and compost them. Stop harvesting when the mature spears only grow to about the diameter of a pencil.

IN THE MARKET

Check the tips of produce-department spears; they should be tightly closed, and the spears should be firm, not limp. The color should be bright. Squeeze a bunch of asparagus and they should squeak lightly if they're fresh. Size doesn't matter. Thicker spears are simply a little more mature than thinner ones.

EXTENDING RIPENESS

Asparagus spears are best cooked as soon after harvest as possible. If you can't use them the day you pick them, wrap the spears in wet paper towels and refrigerate in a vegetable crisper, or store them cut end–down in a glass of cool water. Never store fresh asparagus in a closed plastic bag.

Whether you've bought or picked the asparagus, trim the ends before you use the vegetable because tough fibers form on the ends as a result of harvesting.

For longer term storage, steam or blanch asparagus and refrigerate until you're ready to use it. The asparagus will keep refrigerated for up to a week.

OPTIMAL PLANTING AND HARVEST TIMES		
Asparagus	**Suitable for Succession Planting?**	**N**
Not applicable. Asparagus takes two years to mature. After this, it should be harvested when spears are 7 to 9" tall.		

BASIL

PROLIFIC AND EASY TO GROW, BASIL PROVIDES a continual harvest all season. Harvest regularly to not only collect the leaves, but also to control the growth habit.

"Ripe" is relative with basil. The leaves are just as flavorful small or large. As soon as the plant has at least six leaf pairs, begin harvesting by cutting it back to about half its height. Snip (or pinch) down to just above a leaf pair. This will encourage bushier growth. As the plant matures, snip off leaves as needed. Regularly pinch off the tops of stems to prevent the plant from flowering.

As the weather gets hot, the plant will try to flower. Cut off any flower buds (add them to salads).

The harvest comes to an end at first frost, or when the plant bolts and flowers (once it flowers, the leaves turn unpleasantly bitter). If you're near the end of the season, harvest all the leaves and cut the plant down to the ground.

IN THE MARKET

Basil sprigs are sold in plastic clamshell packaging and fade fast. Inspect the package closely. Reject any with wilting leaves, brown

RIPE. Trim basil flowers down to the second pair of leaves to spur bushy growth and prevent the plant from bolting.

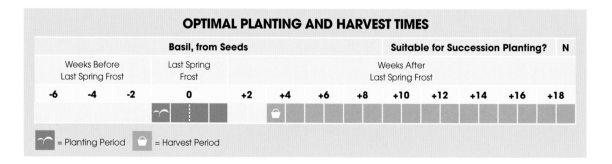

OPTIMAL PLANTING AND HARVEST TIMES

| Basil, from Seeds | | | Suitable for Succession Planting? | N |

Weeks Before Last Spring Frost			Last Spring Frost		Weeks After Last Spring Frost								
-6	-4	-2	0	+2	+4	+6	+8	+10	+12	+14	+16	+18

⤳ = Planting Period   🧺 = Harvest Period

RIPE. Basil at prime cutting time. On all basil, you should regularly pinch off the tops of stems to prevent flowering.

RIPE. Purple basil offers a stunning addition to any dish in which basil is called for. As soon as the plant has at least six leaf pairs, begin harvesting.

spots, or signs of mold. You can also buy packages of "living" basil, increasingly sold in high-end markets. These come enclosed in plastic packaging with the roots intact, and you can store them roots-down in a glass of water to keep the leaves fresh for a week or more.

EXTENDING RIPENESS

Basil lasts less than a week refrigerated. Freeze fresh basil leaves by packing them into a resealable plastic bag or similar airtight container. Dry basil by hanging stems full of leaves in a cool, dark place. Dry individual leaves on large screens. Once dry, crush the leaves and store them in an airtight container in a cool, dark place. Drying concentrates flavor, so use about half the amount of dried basil as a substitute in recipes.

BEANS

BEANS ARE EASY TO GROW AND ARE PROLIFIC garden producers. Most gardeners and cooks feel that pole beans have a superior flavor, but that bush beans produce a more characteristically "green bean" flavor.

Pole beans offer a steady and consistent crop over the season. Pole varieties are great if you prefer to garden vertically due to space constraints, or if you want a reliable supply of beans to cook fresh the day you pick them.

Bush beans produce their harvest all at once, and the beans ripen a little faster—usually about five days earlier—than pole beans do. Bush beans are a good choice for gardeners looking to put up or preserve beans.

No matter which type you cultivate, regularly check your plants. Beans that spend too long on the vine become tough and stringy. The perfectly ripe green bean will be firm, evenly green—not blotchy or yellow—and should have a hydrated, smooth skin. The bean should not be lumpy; if it is, the seeds inside have matured and the bean will be tough and less flavorful.

Carefully harvest ripe beans by pinching off the stem between your thumbnail and forefinger right where the bean connects to the stem. Keep fresh-picked green beans out of direct sunlight and use them as soon as possible.

UNDER-RIPE. Beans are just beginning to form on this decidedly under-ripe string bean plant.

OPTIMAL PLANTING AND HARVEST TIMES

Beans, Bush, from Seeds									Suitable for Succession Planting?			Y
Weeks Before Last Spring Frost			Last Spring Frost				Weeks After Last Spring Frost					
-6	-4	-2	0	+2	+4	+6	+8	+10	+12	+14	+16	+18

Beans, Pole, from Seeds									Suitable for Succession Planting?			N
Weeks Before Last Spring Frost			Last Spring Frost				Weeks After Last Spring Frost					
-6	-4	-2	0	+2	+4	+6	+8	+10	+12	+14	+16	+18

= Planting Period   = Harvest Period

RIPE. Beans at prime ripeness will not yet have pronounced bumps from large inner seeds.

OVERRIPE. At the point where the pods are lumpy because of large inner seeds, green beans will begin to taste tough.

RIPE. Bush beans are better for preservation and have a characteristic "green bean" flavor.

RIPE. Many gardeners and cooks believe pole beans taste better.

IN THE MARKET

Do yourself a favor and buy only loose green beans, because you can't properly inspect the beans if they're packaged. The best green beans in the produce aisle will be slender with no soft spots. They should also be uniformly green; yellowing, bruises, and brown or black spots are bad signs.

EXTENDING RIPENESS

Green beans do not keep well. Store them in the refrigerator in an open plastic bag, and they will last two to four days. If you can't use them immediately, steam or blanch the beans, depending on how you intend to use them. They will keep for up to two weeks if prepared either way. Beans are also excellent vegetables to pickle.

BEETS

BEETS ARE FLAVORFUL AND ENTICINGLY sweet (they've even earned the nickname "nature's candy"), with incredible health-boosting nutrients. Those appealing qualities are all at their heights when the root is perfectly ripe.

The challenge in harvesting this cool-season crop at the perfect moment is that it is hidden underground. Well, half of it. Beet greens are even more nutritious than the root, and many gardeners use the whole plant.

Go by the time to maturity of the variety you're growing, and watch for the shoulders pushing up out of the ground. If you pull your beets up early, they'll be smaller but still flavorful. You'll sacrifice some nutritional value, but the immature beets will be more intensely sweet.

Harvest beet greens when they're about 6 inches long, taking only the outside leaves (shearing all the greens can prevent the beet from maturing). To harvest ripe beets, hold the base of the greens firmly and pull with even pressure. If the beet doesn't easily release from the soil, loosen the surrounding soil with a small garden fork.

IN THE MARKET

When buying beets from a produce bin, look for those that still have the leaves attached. Inspect each beet for any soft spots; the root should be evenly colored and firm all the way around. The greens themselves should still be erect, rather than limp, and appear fresh.

RIPE. Beet greens are just as delectable and nutritious as the roots and can be prepared as you would cook swiss chard.

OPTIMAL PLANTING AND HARVEST TIMES

Beets, from Seeds							Suitable for Succession Planting?					Y
Weeks Before Last Spring Frost			Last Spring Frost		Weeks After Last Spring Frost							
-6	-4	-2	0	+2	+4	+6	+8	+10	+12	+14	+16	+18
	⌢				🧺	▮						

⌢ = Planting Period   🧺 = Harvest Period

RIPE. Harvest beets when root shoulders push up from underground.

EXTENDING RIPENESS

Brush any dirt off your beets as soon as you get them into the kitchen. Cut off the greens, leaving 1 inch on the root (and leaving the tail-like taproot intact). Refrigerate the beets in a vegetable crisper for up to a month. Don't wash them until you're ready to use them. Use the greens right away; they'll usually only keep for two days once cut.

You can store beets over the winter in any cool, dark area (around 35°F). Layer them in damp sand, in wooden boxes or plastic bins with tops. Make sure the beets are not touching. They can last up to five months stored in this way.

BROCCOLI

THE EDIBLE PART OF A BROCCOLI PLANT is the flower, comprised of a cluster of buds. Harvest it at its largest size but before individual buds open.

Check broccoli every day after the head forms. When it is the mature size listed on the seed pack or seedling stick, harvest the head immediately. Cut the stem 4 inches down from the head using a sharp serrated knife. Smaller heads will likely grow offshoots from the main stem for a second crop.

IN THE MARKET

Broccoli head buds should be tightly closed and the head should have a compact structure. Browned stem ends are signs of age. The head should be evenly green with no yellow spots or buds, and it should be heavy for its size.

EXTENDING RIPENESS

Eat broccoli as soon as possible after harvest. A head of broccoli will keep for a week if refrigerated.

UNDER-RIPE. Immature broccoli heads have irregular-sized buds. The broccoli won't be ripe until the buds are closed tight and are uniform.

RIPE. Harvest broccoli before individual buds open.

OPTIMAL PLANTING AND HARVEST TIMES

Broccoli, from Seedlings												Suitable for Succession Planting?	Y
Weeks Before Last Spring Frost			Last Spring Frost					Weeks After Last Spring Frost					
-6	-4	-2	0	+2	+4	+6	+8	+10	+12	+14	+16	+18	
⌢			⋮		🧺								

⌢ = Planting Period   🧺 = Harvest Period

BRUSSELS SPROUTS

BRUSSELS SPROUTS BEGIN FORMING ALONG the bottom of the stem, developing further up as the plant grows and matures over the course of a few weeks. The first sprouts can be harvested when they are tight and firm and between 1 and 2 inches in diameter. Simply twist off ripe sprouts. It should take very little effort.

If you're nearing winter and want all the sprouts to mature at once, cut off the top of the plant. All the sprouts will ripen within three weeks.

IN THE MARKET

Carefully pick individual sprouts that are heavy for their size and that have a vibrant green color. Each sprout should be firm and compact with tightly wrapped leaves. Avoid any with yellowing leaves or brown stem ends, or those that are soft and loose. Size won't affect the flavor.

EXTENDING RIPENESS

Don't wash sprouts before you store them. Instead, place them in an unsealed plastic bag and refrigerate. They'll keep for up to five days, but the flavor is at at its best the sooner you eat them after harvest or purchase.

RIPE. Brussels sprouts can be harvested when they are tight, firm, and 1 to 2" in diameter.

OPTIMAL PLANTING AND HARVEST TIMES

Brussels Sprouts, from Transplants													Suitable for Succession Planting?	N
Weeks Before Last Spring Frost			Last Spring Frost	Weeks After Last Spring Frost										
-6	-4	-2	0	+2	+4	+6	+8	+10	+12	+14	+16	+18		

= Planting Period   = Harvest Period

CABBAGE

CABBAGE IS A COOL-SEASON CROP THAT must have a reliable supply of water, adequate nutrients, and careful tending. All types of cabbage share traits, but there are differences.

NAPA

This delicate cabbage has a fresher, brighter flavor than green cabbage. Napa forms a cylindrical head that looks like a cream-and-green barrel. Determine ripeness by squeezing the head. It's ready to harvest when the leaves are tight and your grip meets resistance. Napa cabbage will withstand a light frost. Harvest the head, strip the outer leaves, loosely wrap in an unsealed plastic bag, and thaw in the refrigerator overnight.

GREEN OR RED

Green or red cabbage that looks ripe may still be maturing. A ripe head feels solid, and the leaves will be firmly wrapped around the head. Any looseness is a sign of immaturity. Red cabbage is smaller and denser than green. Harvest either by cutting the head at the base using a sharp serrated knife. Leave the loose outer leaves attached, and the plant may grow small heads for a second harvest.

SAVOY

This crinkle-leaf cabbage is harvested in the same way as green cabbage. Savoy's leaves make visual inspection difficult. Once the head begins to grow, feel the head. When ripe, it will be firm with a slight give due to the wrinkles creating air spaces.

IN THE MARKET

Look for solid, hefty heads that are surprisingly heavy. The color should be vibrant, and the leaves crisp, not limp. Avoid heads with leaf insect damage or those that have split. Most cabbage is available year round, but the sweetest cabbage will be found in early winter.

OPTIMAL PLANTING AND HARVEST TIMES

Cabbage, from Seedlings													Suitable for Succession Planting?	Y
Weeks Before Last Spring Frost			Last Spring Frost	Weeks After Last Spring Frost										
-6	-4	-2	0	+2	+4	+6	+8	+10	+12	+14	+16	+18		

= Planting Period   = Harvest Period

RIPE. Cabbage is ready for harvest when heads are tight and solid.

EXTENDING RIPENESS

Store whole cabbage heads, wrapped loosely in a plastic bag, in the refrigerator. They will keep for a week. Refrigerate cut heads with the cut face wrapped tightly in plastic wrap. Cut cabbage will keep for several days. Remove the outer leaves and thoroughly wash before using the cabbage.

Cabbage is also ideal for root cellar storage. When the head is mature, dig up the entire plant, lay it out on a bed of straw in a row with other cabbage plants, and separate and top each plant with a healthy layer of straw. The cabbage will last two months or more.

THE 10 HEALTHIEST RIPE FRUITS AND VEGETABLES

Ripening is a plant's way of peaking nutritional value. If health is your prime consideration, consider these garden standouts.

1. Kale

Kale is a "superfood" nutritional powerhouse. Low calories, no fat, more iron than beef—what's not to love? Kale offers abundant vitamin K—a cancer-prevention nutrient—and vitamins A and C. It is rich in antioxidants and anti-inflammatory compounds, and has been proven to reduce cholesterol.

2. Broccoli

Broccoli is a great source of fiber, with a unique combination of vitamins A and K (thought to promote production of essential vitamin D that is usually produced through exposure to sunlight). The vegetable has excellent allergy-fighting compounds.

3. Pomegranates

Pomegranate juice has been found to have a greater amount of antioxidants than red wine, cranberry juice, or even green tea.

4. Brussels Sprouts

When it comes to glucosinolates, you can't top brussels sprouts. Glucosinolates are phytonutrients that set the stage for cancer prevention in our bodies and our cells.

5. Tomatoes

Aside from abundant vitamin C, tomatoes have significant levels of an antioxidant called lycopene. A disease-fighter, lycopene may help maintain bone health, and works with other compounds to fight heart disease.

6. Beets

Beets contain special phytonutrients called betalains—antioxidants, anti-inflammatories, and detoxifiers. The greens are even more nutritious, containing whopping amounts of lutein, a compound associated with eye health.

7. Collard Greens

Collard greens are chock-full of folate, calcium, vitamins K and C, and beta carotene. The cooked greens also help lower cholesterol.

8. Summer Squash

Summer squash boasts strong antioxidants including carotenoids, lutein, zeaxanthin, and others. Certain compounds in summer squash may play a role in diabetes prevention. The skin is rich in nutrients, so prepare the squash with the skin left on.

9. Blueberries

New studies show that ripe blueberries play an important role in improving memory and could slow down cognitive decline as a part of aging.

10. Carrots

Carrots grown in season are likely to be the most nutritious. Ripe carrots are most well-known for high levels of carotenoids, but they may also play a role in preventing cardiovascular disease.

CANTALOUPE

GARDEN-GROWN CANTALOUPE IS SWEETER than most you'll find in grocery stores, and delivers abundant amounts of vitamins A and C. The ripest, most flavorful cantaloupe requires a good start in rich, loamy soil that is already sun warmed. Growing cantaloupe requires about 2 inches of water a week. Reduce watering once the fruits are set and growing to intensify sweetness.

When ripe, the cantaloupe's skin turns from smooth and green to a rough "netting" texture. The underlying color will turn yellow to tan.

When ready to harvest, a crack may develop in the stem where it attaches to the fruit. A ripe cantaloupe will feel heavier than it looks and will have a distinctive, floral, musky-sweet smell. A ripe cantaloupe will easily separate from the stem with a light twist.

Be aware that all the melons on the vine will ripen over a few short weeks. Check them daily to protect against overripening and spoilage.

IN THE MARKET

The more pronounced the floral, musky-sweet smell is, the more certain you can be that the cantaloupe is ripe. The color behind the rough netting surface texture should be golden or buff yellow. If it's green, the melon was picked early for transport to market.

Age an unripe cantaloupe on the counter for a few days. The aging will not noticeably increase the sweetness, but it will soften the flesh to a riper texture.

UNDER-RIPE. Unripe cantaloupe remains smooth and green.

OPTIMAL PLANTING AND HARVEST TIMES

Cantaloupe, from Seedlings													Suitable for Succession Planting?	N
Weeks Before Last Spring Frost			Last Spring Frost	Weeks After Last Spring Frost										
-6	-4	-2	0	+2	+4	+6	+8	+10	+12	+14	+16	+18		
				⌒				🧺						

🗺 = Planting Period   🧺 = Harvest Period

RIPE. The rough netting texture and tan coloring of a ripe cantaloupe is readily apparent. A ripe cantaloupe feels heavier than it looks.

EXTENDING RIPENESS

Store whole cantaloupes in the refrigerator for about a week. Tightly wrap a cut melon in plastic wrap and it will last for about three days in the refrigerator. To enjoy the fruit for much longer, slice the cantaloupe and cut the rind off. Chop the flesh into cubes and freeze them. The flavor and sweetness will hold up for weeks when frozen, and you can use the chunks in everything from smoothies to sweet cantaloupe soup.

GARDEN WISDOM: DON'T PINCH!

Conventional wisdom holds that it's smart to pinch off a cantaloupe vine's shoots so that the plant will divert energy to the fruits. Research shows the folly of this practice. The foliage helps the fruit produce sugar—less foliage means less sugar.

CARROTS

ALTHOUGH CARROTS BECOME SOMEWHAT tough if you leave them in the ground for too long, there is a lot of latitude when it comes to harvesting carrots.

UNDER-RIPE. Under-ripe carrots' shoulders aren't yet showing through the ground.

Use the maturation time listed on the seed packet as a baseline. Carrots will also give you clear indicators of ripeness. The shoulders will begin pushing up out of the ground, revealing the carrot's diameter. If you don't see the shoulders and you suspect the carrots are mature, lightly dig out the soil around one carrot to determine diameter. While ¾ to 1 inch is standard, that measurement varies depending on the variety.

Err on the side of an early harvest. Small carrots (in contrast to specific miniature carrot varieties grown as baby carrots), have a lighter, more delicate flavor.

Harvesting is the same regardless of varietal. Grab the base of the greens and slowly and evenly pull the carrot out of the soil. If it is stubborn, gently loosen the soil, being careful not to touch the carrot. You can also moisten the soil.

IN THE MARKET

Bunches of carrots with greens attached are more expensive and the greens can cause moisture loss. Loose carrots are easier to inspect. They should be uniform shapes and firm. Avoid any with rootlets attached, or with noticeable cracks.

OPTIMAL PLANTING AND HARVEST TIMES

Carrots, from Seeds													Suitable for Succession Planting?	Y
Weeks Before Last Spring Frost			Last Spring Frost				Weeks After Last Spring Frost							
-6	-4	-2	0	+2	+4	+6	+8	+10	+12	+14	+16	+18		
	〜					🧺								

〜 = Planting Period   🧺 = Harvest Period

RIPE. This dwarf variety of carrot is ripe. The diameter of the carrot to determine ripeness differs by variety.

Mass-farmed carrots are often bunched in bags with bright orange designs. These can trick the eye and make it hard to judge the carrots inside. Healthy, ripe carrots are a uniform color top to bottom (the darker, the better). The shoulders can be slightly green, but they should not be darkened or black.

EXTENDING RIPENESS

Twist off the carrot greens, leaving about an inch of stem. Eat the greens within a day or two because the quality deteriorates quickly. To store the carrots in the refrigerator, rinse them and refrigerate in an unsealed plastic bag. They should stay firm and flavorful for up to a week.

Keep carrots for longer periods by twisting off the tops, brushing off any dirt, and layering them in a tub or large box with damp sand. The individual carrots should not touch, and you should cover them with a few inches of straw. Store them all winter in a cool, humid area at around 35°F.

CAULIFLOWER

CAULIFLOWER PLANTS NEED AROUND SIX HOURS of sun a day, but the head has to be shaded or else it will become discolored and bitter (unless the variety you're growing is self-blanching). Blanch cauliflower by wrapping and tying the leaves around the head. Most varieties ripen two weeks after blanching starts, but timing varies and doesn't apply to self-blanching varieties. Mature heads are 6 to 8 inches in diameter, firm, and compact, but check seed packets or seedling sticks for your cultivar's mature diameter.

Harvest early and flavor will be under-developed; too late and the head will start separating and opening, the texture becoming

UNDER-RIPE. This cauliflower is too small; the head is just emerging.

OPTIMAL PLANTING AND HARVEST TIMES

Cauliflower, from Transplants												Suitable for Succession Planting?	N
Weeks Before Last Spring Frost			Last Spring Frost	Weeks After Last Spring Frost									
-6	-4	-2	0	+2	+4	+6	+8	+10	+12	+14	+16	+18	

= Planting Period   = Harvest Period

RIPE. Mature cauliflower heads are 6 to 8" wide, firm, and compact.

coarse, grainy, and unpalatable. If the head begins to open, harvest immediately.

Use a sharp knife to cut a ripe head from the stem at soil level, leaving the base leaves attached to the head (they'll protect the cauliflower until you use it). The stem may grow other florets for a second harvest.

IN THE MARKET

Buy cauliflower heads that are a bright, unblemished white. There should be no spots or deformities. The larger the leaves, the better.

EXTENDING RIPENESS

A full head of cauliflower will keep one to two weeks refrigerated in a plastic bag. Precut florets will only last a few days.

CELERY

CELERY LEFT LOOSE WILL SPREAD AS IT grows. Stalks mature deep green, nutritious, and fibrous. Blanch celery by tying the bunch tightly together. The outer stalks shield the inside, ensuring that inner stalks stay pale and tender. Harvest individual stalks as the plant grows, anytime they're more than 6 inches tall. Cut stalks 2 inches up from the soil with a sharp knife, or harvest the entire bunch when it is 3 to 4 inches in diameter. Use a sharp knife to slice through the base at soil level.

IN THE MARKET

It's wise to buy organic celery because celery sucks up pesticides and chemicals like a sponge. The celery should be firm and crisp with a light, fresh smell. Avoid bunches with blemishes, signs of insects, limpness, or a musty smell.

EXTENDING RIPENESS

Refrigerate celery in a loose plastic bag. Single stalks go limp more quickly, but even a bunch will not last more than five days. Store an entire fall crop in a root cellar or a basement. Dig the plants up with the roots and pack them in moist sand in tubs. Kept at 35°F, the celery will last two to three months.

∧ **UNDER-RIPE.** A healthy celery plant takes root but is not yet large enough to be ripe.

> **RIPE.** Celery stalks can be harvested as soon as they reach 6" tall. Harvest individual ripe stalks as the plant grows.

OPTIMAL PLANTING AND HARVEST TIMES

Celery, from Transplants									Suitable for Succession Planting?		N	
Weeks Before Last Spring Frost			Last Spring Frost		Weeks After Last Spring Frost							
-6	-4	-2	0	+2	+4	+6	+8	+10	+12	+14	+16	+18

⌢ = Planting Period   ⛏ = Harvest Period

CHIVES

CHIVES PRODUCE AN ONGOING HARVEST OF delicately flavored leaves. The oniony flavor is the same regardless of size. Start harvesting as soon as the plant is 3 inches tall. Cut the leaves down to 1 inch above the soil with sharp scissors. The plant will grow back, and you can expect at least three harvests in a season. Harvest the flowers to add to soups and salads.

IN THE MARKET
Chives are sold in plastic clamshell packaging, or as living herbs in small, disposable pots. The chives should be a uniform green with no yellowing or blemishes. Check for condensation and that no part of the chives is slimy and wet. Use packaged chives immediately. If you're buying living herbs, look for young chives and avoid pots that are overcrowded.

EXTENDING RIPENESS
Chives are best used the day you pick or buy them because they're delicate and lose flavor quickly, even when refrigerated. You can freeze chives, but don't dry them because they lose their flavor when dried. If chives are a favorite herb, grow them on a sunny windowsill.

RIPE. When the chives plant is 3" tall, cut the leaves down to 1" to harvest.

OVERRIPE. Although chives flower at the end of the season, the flowers can be used in soups and salads.

OPTIMAL PLANTING AND HARVEST TIMES													
Chives, from Transplants								Suitable for Succession Planting?					Y
Weeks Before Last Spring Frost			Last Spring Frost		Weeks After Last Spring Frost								
-6	-4	-2	0	+2	+4	+6	+8	+10	+12	+14	+16	+18
			🌱		🧺								

🌱 = Planting Period   🧺 = Harvest Period

CILANTRO

CILANTRO LEAVES CAN BE HARVESTED ONCE the plant reaches 6 to 8 inches tall; regularly trimming foliage prevents flowering. But flowering doesn't signal the end of the harvest. Flowerheads provide seeds—known as coriander seeds—that can be used as a spice. Cut off outside leaves every week using sharp scissors. Cut off mature flowerheads and place them upside down in paper bags. When they dry, the husks split and drop the seeds.

IN THE MARKET

Cilantro bunches should be a uniform vibrant green with firm stems and no signs of wilting.

Fresh cilantro is fragrant with citrus tones. Avoid any with even a few yellow or brown leaves.

EXTENDING RIPENESS

Cilantro is best used the day it's harvested. It will keep for several days refrigerated in a plastic bag, and even longer if the fresh-cut stems are put in a glass of cool water and covered with a plastic bag. Drying destroys the distinctive flavor, but some cooks freeze chopped cilantro in water in ice cube trays for soups and sauces. Coriander seeds should be stored in a closed container in a cool, dark space.

RIPE. Cilantro is ripe when it reaches 6 to 8" tall.

OVERRIPE. The seeds within the flowerheads on cilantro are known as coriander—a desired spice.

OPTIMAL PLANTING AND HARVEST TIMES

Cilantro, from Seeds or Transplants												Suitable for Succession Planting?	Y
Weeks Before Last Spring Frost			Last Spring Frost		Weeks After Last Spring Frost								
-6	-4	-2	0	+2	+4	+6	+8	+10	+12	+14	+16	+18	

= Planting Period   = Harvest Period

COLLARD GREENS

THIS COOL-WEATHER SOUTHERN FAVORITE usually matures within about two-and-a-half to three months, after which the lower leaves can be harvested and eaten. Judge ripeness by size. Harvest when the plant reaches about 1 foot tall. Cut individual leaves from the bottom up, before they get too old and become tough—ideally, while they are less than 8 inches long. Simply break the leaves off the main stem to harvest them.

Although you can grow collard greens in spring or fall, fall varieties can be slightly more rewarding because a light frost makes the flavor noticeably sweeter. When the weather heats up or the first hard freeze is on the way, harvest all the remaining leaves, dig the plant up, and compost it.

IN THE MARKET

Collard greens do not travel well, so look for dark green, firm leaves from a local supplier. Soil sticks to the leaves and a thorough cleaning is part of the kitchen prep necessary to cook collard greens. However, avoid any yellowed or wilted leaves, or those with tears or holes.

UNDER-RIPE. Under-ripe collard greens look good but need to grow more for peak flavor.

OPTIMAL PLANTING AND HARVEST TIMES

Collard Greens, from Seeds				Suitable for Succession Planting?		N
Weeks Before Last Spring Frost			Last Spring Frost	Weeks After Last Spring Frost		

-6	-4	-2	0	+2	+4	+6	+8	+10	+12	+14	+16	+18

〰 = Planting Period   🧺 = Harvest Period

RIPE. Simply break the leaves off the main stem to harvest ripe collard greens.

EXTENDING RIPENESS

You can simply store harvested collard greens in plastic bags in the refrigerator, where they'll last for five to seven days. However, a quick soak in ice water is said to preserve the greens' freshness. Lightly dry the leaves and store them in loose plastic bags in a vegetable crisper drawer. You can also store the leaves wrapped in moist paper towels and put them in an unsealed plastic bag before refrigerating.

Diligently wash the leaves before using them if you want to avoid grit in your collard green recipes. Soak the leaves in a bowl of lukewarm water and change the water until no more dirt comes off the leaves.

You can also freeze collard greens. Wash the greens thoroughly until you're sure there is no grit remaining, then plunge the leaves into boiling water for three minutes, transfer to an ice bath until chilled, and drain. Put the leaves in large resealable bags with the air squeezed out, and the leaves should keep up to a year in the freezer.

CORN

PERFECTLY RIPE SWEET CORN IS A FLEETING summer treat; you'll be lucky to get two ears from a single stalk. But get the timing spot-on and few vegetables taste as good.

As the corn matures, the plant produces abundant sugars to fuel growth. But once the ears are mature, the sugar inside the kernels quickly converts to less flavorful starch. That change proceeds even after you harvest the ear. You should ideally cook the corn the day you pick it. (Almost half the sugar in the kernels will be converted into starch within the first day after harvest.)

Make sure you know the maturation period for the variety you've planted. Timing ranges from 60 days to more than 85 days for "super sweet" types. But it's more important to feel the ears as they ripen. Ripe corn is rounded at the top, and the ear will feel plump and full. Unripe ears are skinny and pointed. Ripeness is somewhat a matter of taste. Some people find very full, plump kernels to be overripe.

The silk—the soft, velvety strands sticking out of the top of the ear—on a ripe ear will be dried out and brown, with a little green where it comes out of the ear. It will also be sticky from the pollen it has collected.

Harvest ripe corn in the morning. Hold the stalk with one hand and firmly pull the ear down with the other, twisting it slightly to break the connection to the stalk.

UNDER-RIPE. Unripe ears are skinny and pointed. Ripe corn is rounded at the top, and the ear will feel plump and full.

OPTIMAL PLANTING AND HARVEST TIMES

Corn, from Seeds							Suitable for Succession Planting?				Y
Weeks Before Last Spring Frost			Last Spring Frost		Weeks After Last Spring Frost						
-6	-4	-2	0	+2	+4	+6	+8	+10	+12	+14	+16	+18

⌒ = Planting Period   ⛑ = Harvest Period

IN THE MARKET

Buying the freshest corn possible is key to enjoying the best texture and sweetest flavor. Roadside stands and farmers' markets are often the best bet, because the corn has often come out of the fields that day. If you're buying your corn at the local supermarket, check that the husk is intact and topped with silk. Don't buy shucked corn. It may save you some labor, but it's likely far from fresh.

The silk should be brown but still sticky. Silk that is entirely dry to the touch or black rather than brown is a sign that the ear is past its prime. You can pull down the husk slightly to check that the kernels are plump and bright, and that they are so full that they will puncture with slight pressure. If you're looking for a particular color of corn (yellow, white, or bicolor), ask the produce manager or the farm stand clerk.

EXTENDING RIPENESS

Refrigerate ears in their husks as soon as possible. This arrests the conversion of sugar to starch, but doesn't stop it completely. The sweetness and texture of corn will continue to degrade rapidly over time. The corn will keep in the refrigerator for five to seven days.

If you want to preserve the corn for future use, blanch it on or off the cob, then freeze it. Frozen corn will keep for six months or more. You can extend your crop of ripe corn by succession planting. Learn more about that strategy in Chapter 3, page 157.

RIPE. Tassels that have turned from white to brown indicate a cob is ripe. Ripe corn will feel plump and full.

CUCUMBERS

THE TWO BASIC TYPES OF CUCUMBER RIPEN and are harvested in the same manner. **Slicing cucumbers** are more flavorful and are meant to be eaten raw. **Pickling cucumbers** are uglier with bumpy skin. They are blanched and pickled.

Cucumbers peak about ten days after the first female flowers open. Know the mature size of your variety and err on the side of picking the cucumbers slightly small. Harvest in the morning, using garden shears to cut the stem. Handle harvested cucumbers carefully to avoid bruising.

IN THE MARKET

Fresh cucumbers are firm and uniformly dark green. If you eat the skin, make sure the cucumbers haven't been waxed. Smaller cucumbers are less likely to be bitter. Avoid any with even small bruises or yellow spots.

English cucumbers are longer greenhouse fruit with a pronounced floral scent and delicate flavor. They are sold wrapped in plastic to prevent dehydration (the thin skin is usually eaten). Avoid any with even tiny soft spots.

EXTENDING RIPENESS

Ideally, eat the cucumbers the day they come off the vine or from the store. Refrigerate cucumbers in a loose plastic bag for up to five days.

∧ **UNDER-RIPE.** Under-ripe cucumbers are undersized; this variety will be ready for harvest once it is 5 to 8" in length.

> **RIPE.** Cucumbers peak 50 to 70 days after planting. Ripe cucumbers will have smooth skin and will be uniformly dark green.

OPTIMAL PLANTING AND HARVEST TIMES

Cucumber, from Transplants													Suitable for Succession Planting?	N
Weeks Before Last Spring Frost			Last Spring Frost		Weeks After Last Spring Frost									
-6	-4	-2	0	+2	+4	+6	+8	+10	+12	+14	+16	+18		

= Planting Period ⌒   = Harvest Period ⬒

EGGPLANT

REGARDLESS OF SHAPE OR MATURE COLOR, eggplants are all harvested the same way. It's best to pick them slightly early, because overripe eggplants are bitter. Judge ripeness by the mature size for the variety you're growing. Once the fruit is close to mature size, the color is set, and the skin is glossy, it's ready to be picked.

Wear gloves because eggplant stems have sharp prickles. Cut the stem 1 inch up from the fruit with sharp pruning shears. Regular harvesting can boost yield.

IN THE MARKET

Ripe eggplants have taut, firm, and shiny skins. The eggplant should feel slightly heavy for its size. Brown spots, discoloration, scars, or obvious bruising are signs of mishandling. The freshest eggplants have healthy green caps and stems, with no browning or mold.

RIPE. As soon as the skin is glossy, an eggplant is ready to be picked, regardless of size. Waiting until the fruit deepens to black may result in a bitter eggplant.

EXTENDING RIPENESS

Rinse newly harvested eggplant and pat dry with a paper towel. Refrigerate for up to five days. Many experts, however, believe that the eggplant will last just as long stored in a dark, cool, dry location on a counter or in a cabinet. Cut eggplants will last two to three days refrigerated and wrapped in plastic wrap

OPTIMAL PLANTING AND HARVEST TIMES

Eggplant, from Transplants								Suitable for Succession Planting?				N
Weeks Before Last Spring Frost			Last Spring Frost	Weeks After Last Spring Frost									
-6	-4	-2	0	+2	+4	+6	+8	+10	+12	+14	+16	+18	

= Planting Period   = Harvest Period

RIPE. Taught, firm, and shiny—this eggplant is ready for harvest.

OVERRIPE. Left a little too long on the vine, an eggplant skin will begin to toughen and split.

GARLIC

HARVEST GARLIC TOO EARLY AND BULBS WON'T keep. Too late and the protective covering will be compromised. That's true of both of the two basic types. **Softneck** are more popular because they have thicker "wrappers," the papery covering around the individual cloves. **Hardneck** varieties have a hard central stem and a milder flavor.

The greens die as the bulb matures. When they first start dying, stop watering for a week. Harvest when the leaves on the bottom half of the stem are all brown.

On a dry day, loosen the soil with a garden fork, being careful not to touch the bulb. Dig under the bulb and force it to the surface intact. Immediately move the bulb into shade or into the kitchen—it quickly blanches in direct sunlight.

IN THE MARKET

Squeeze the garlic bulb. It should be firm, without dried or soft cloves. Reject heads with sprouting cloves. The cloves should be covered by wrappers so that no flesh is exposed to the air.

UNDER-RIPE. The greens die as the garlic bulb matures.

EXTENDING RIPENESS

Cure fresh-picked garlic if you're not using it immediately. Leave leaves and roots intact. Gently clean off any dirt stuck to the bulb. Spread bulbs on drying screens and cure for two weeks in a cool, dry, well-ventilated location such as a shed or garage. When the papery covering and roots are entirely dry, trim the bulbs, cutting down the stem of hardneck varieties. Store in a cool, dry place out of direct sunlight—never in the refrigerator.

OPTIMAL PLANTING AND HARVEST TIMES		
Garlic	**Suitable for Succession Planting?**	**Y**
Too variable to be timed. Garlic is usually planted in the fall for harvest the next spring and summer.		

RIPE. To harvest, loosen the soil with a garden fork, being careful not to touch the bulb. Dig under the bulb and force it to the surface intact.

KALE

HARVEST KALE LEAVES AS THEY MATURE, OR pick more tender, milder young leaves. There are several interesting varieties in addition to common curly kale, which you can grow or look for at market.

Lacinato kale. The deep blue-green leaves are shaped like feathers and have deeply textured surfaces. It's sweeter and nuttier than curly kale.

Red kale. Sometimes called red Russian kale, this type has red stems, green frilly leaves, and a milder flavor than curly kale, with undertones of pepper.

Redbor kale. This kale is a fuchsia-red all over, with frilled leaves. It has a traditional kale flavor and is sometimes grown as an ornamental.

Kale leaves can be harvested any time after they are 6 to 8 inches long (compost any leaves that are damaged or yellowed). Harvest leaves from the bottom of the stem, leaving those above to mature and spur new growth. There should always be three to four leaves at the crown.

You can break the leaves off the main stem, but it's safer to cut them where they grow from the stem. Harvest the entire plant when it's about to bolt—or when the temperatures are going to fall below 20°F—by cutting the main stem at soil level.

IN THE MARKET

Check bunches of kale for any signs of leaf damage or yellowing. Look for smaller leaf bunches to take advantage of the lighter, more tender leaves. Kale should never be limp or dull. The stems should be firm but pliable. It's wise to buy organic, because kale has been shown to absorb chemical contaminants.

EXTENDING RIPENESS

Refrigerate fresh kale in a loose plastic bag. It will keep for up to a week, although it becomes more bitter with age. Wash the leaves thoroughly before using because kale leaves trap dirt.

Some cooks freeze kale for use in cooked dishes later on. You can try this by cleaning the kale, chopping it into large pieces, and spreading them on small baking sheets to go into the freezer. Once the pieces are frozen, transfer them to a resealable bag. Thaw them out to substitute them for use in cooked recipes calling for fresh kale.

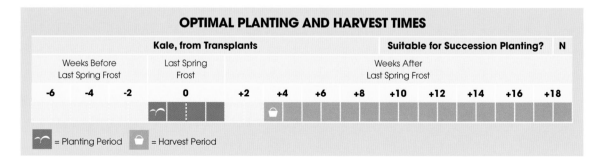

OPTIMAL PLANTING AND HARVEST TIMES

Kale, from Transplants						Suitable for Succession Planting?		N

Weeks Before Last Spring Frost			Last Spring Frost		Weeks After Last Spring Frost								
-6	-4	-2	0	+2	+4	+6	+8	+10	+12	+14	+16	+18

= Planting Period   = Harvest Period

RIPE. Curly kale is ripe from the moment full leaves appear, and does not ever become overripe. Kale leaves can be harvested any time after they are 6 to 8" long.

RIPE. Lacinato kale has deep blue-green leaves.

RIPE. Red stems and green frilly leaves are the hallmark of ripe red kale.

RIPE. Sometimes grown as an ornamental, redbor kale has frilled, fuchsia-red leaves.

LOOSE-LEAF LETTUCE

THIS IS A FORGIVING CROP WHEN IT COMES to determining ripeness. As a cool-season crop, the plant keeps growing as long as the weather isn't hot (consistently above 75°F). If you harvest older outer leaves, the plant will grow more inner leaves, providing an ongoing crop until the plant finally bolts.

Leaf lettuces are some of the most popular picks for home gardeners precisely because they are so prolific. A fall crop will produce past the first light frost, and will even taste sweeter because of it.

Although there are several different types and many varieties of loose-leaf lettuces, they are all harvested in the same fashion. The only wrong way to harvest any of these is to allow the plant to mature to the point of bolting—sending up a seed stalk—at which point the leaves become unpalatably bitter.

Cut off outside leaves at any point when they're more than 3 inches tall (regardless of the listed mature size for the variety you're growing) using a pair of sharp scissors. You can snap off the leaves if you prefer, but be careful to avoid damaging the plant in the process. Harvest in the cool of the morning to keep the leaves as crisp as possible and get them inside as quickly as possible. Once the weather heats up and the lettuce prepares to bolt, cut the entire head at the base of the plant.

UNDER-RIPE. Under-ripe loose-leaf lettuce will keep growing as long as the weather isn't too hot.

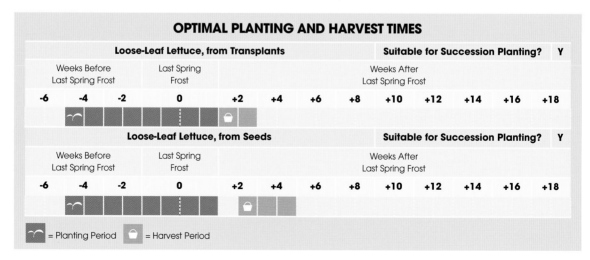

OPTIMAL PLANTING AND HARVEST TIMES

Loose-Leaf Lettuce, from Transplants											Suitable for Succession Planting?	Y

Weeks Before Last Spring Frost			Last Spring Frost					Weeks After Last Spring Frost					
-6	-4	-2	0	+2	+4	+6	+8	+10	+12	+14	+16	+18	

Loose-Leaf Lettuce, from Seeds											Suitable for Succession Planting?	Y

Weeks Before Last Spring Frost			Last Spring Frost					Weeks After Last Spring Frost					
-6	-4	-2	0	+2	+4	+6	+8	+10	+12	+14	+16	+18	

〜 = Planting Period   ⌂ = Harvest Period

RIPE. Loose-leaf lettuce remains edible as long as leaves are harvested regularly. Cut off outside leaves at any point once they're more than 3" tall.

OVERRIPE. With this thick central stalk, this plant is about to "bolt" and flower. It now will be rather bitter.

IN THE MARKET

Although most loose-leaf lettuce offered in large supermarkets has been bagged, it's usually wiser to look for unbagged lettuce so that you can fully inspect the leaves. The processing necessary to deliver bagged lettuce can also steal nutrients and add time between harvest and market, which translates to deteriorating lettuce quality. Look for loose bins of lettuce from local sources for the freshest loose-leaf lettuce. The leaves should appear well hydrated and vibrant. Avoid limp or dull leaves.

EXTENDING RIPENESS

Rinse newly harvested lettuce leaves and then dry them in a salad spinner or pat dry with paper towels. Refrigerate the dry leaves in a resealable plastic bag. They'll keep for three to five days, but are crispest when used fresh.

LEEKS

THERE ARE FAST-MATURING SUMMER OR autumn leeks, and overwintering varieties that take longer to mature and can withstand cold. They're harvested in the same way, and can be pulled up as soon as they grow to a usable size. But most gardeners wait until the leek is more than 1 inch in diameter. Cut one or two young leaves to flavor soups or stews as the leek grows, leaving the rest so that the leek will still grow to maturity.

Let short-season varieties grow until the first frost, or until the weather turns hot. Overwintering leeks can be left in the ground until the first hard freeze, or you can even leave them over the colder months in warmer zones of the country. Spring leeks should be harvested before they bolt and send up their flower stalks. Once bolted, the leek will lose its tenderness and be stringy and unappetizing. To harvest any leek, carefully loosen the soil around the vegetable and pull it up.

UNDER-RIPE. The nearly translucent green leaves of under-ripe leeks.

IN THE MARKET

Shop for leeks with pure white bottoms and stiff green leaves on top. The roots should still be attached. Avoid leeks with blemishes or gashes on the stalk, or damage to the leaves. Be aware that leeks approaching 2 inches in diameter may have tough, woody cores. Smaller leeks will be more tender.

OPTIMAL PLANTING AND HARVEST TIMES

Leeks, from Transplants									Suitable for Succession Planting?			Y
Weeks Before Last Spring Frost			Last Spring Frost	Weeks After Last Spring Frost								
-6	-4	-2	0	+2	+4	+6	+8	+10	+12	+14	+16	+18

= Planting Period   = Harvest Period

RIPE. Most gardeners wait until the leek is more than 1" in diameter to harvest leaves.

EXTENDING RIPENESS

Use leeks as fresh as possible. Don't wash them if storing. Chop off the top of the leaves and wrap the leeks loosely in a plastic bag before refrigerating for up to a week. Smaller leeks have a longer shelf life, so use the largest leeks first.

Remove all grit from leeks by trimming the roots and dark green portions of the leaves. Cut the stalk in half lengthwise, rinse, and then submerge in a large bowl of clean water, cut-side down. Empty the bowl and repeat with clean water until the water is clear after the leek halves have been swished around. Let the leek halves dry before using.

You can also store leeks in a root cellar by leaving the roots on and cutting the leaves back to 1 inch above the white. Pack them standing upright and not touching in a box filled with moist sand. They'll keep for up to six months.

MINT

YOUNGER MINT LEAVES ARE MORE FLAVORFUL than older ones, and regular trimming keeps the plant productive. For a bumper crop, cut the plant down to 1 to 2 inches above the soil right before it begins flowering (when the flavor is most intense). The plant will regrow, and you can repeat the process two to three times in a season. Harvest individual leaves by pinching them off at the stem between your thumbnail and forefinger.

IN THE MARKET

Mint is most often sold as sprigs in plastic clamshell packaging. Look for perky, bright green leaves. Avoid any mint with limp leaves, apparent rotting in the stems, or leaf blemishes such as brown spots.

EXTENDING RIPENESS

Mint sprigs with roots attached can be kept in a glass of water as if they were cut flowers. They'll stay fresh for a week or more. Keep fresh-picked mint leaves in a plastic bag in the refrigerator for two to three days, or dry your mint leaves on a screen in a cool, dark, well-ventilated area. Dried mint will keep for up to a year and can be used for mint tea.

RIPE. Mint leaves are best harvested when young.

UNDER-RIPE. Although this young mint plant is under-ripe, it has very flavorful leaves.

OPTIMAL PLANTING AND HARVEST TIMES

Mint	Suitable for Succession Planting?	N
Not applicable. Mint is a perennial herb; it can be harvested all season.		

OKRA

OKRA IS A QUICK-GROWING, HOT-WEATHER plant common in the South. It's ready for harvest two months after the seed is planted. Watch the plant closely on a daily basis once it flowers. The okra is usually ripe four to five days later, when pods are 2 to 3 inches long.

Wear long sleeves and gloves because the plant grows tiny spines that can poke and irritate the skin. Cut the stem right above the ripe pod's cap, using a sharp knife.

If the stem seems impossible to cut, the pod has overripened. Use shears to remove and discard it.

IN THE MARKET

Ripe, fresh okra is bright green (some varieties are purple, and the color should be fresh and vibrant), and firm but not hard. No pod should be longer than 4 inches. Avoid soft or limp pods, or any with obvious cuts or blemishes.

EXTENDING RIPENESS

Okra is best fresh, but you can store it in the refrigerator crisper drawer in a loose plastic bag for about a week. You can also freeze okra in a resealable plastic bag.

UNDER-RIPE. Under-ripe okra has pods not quite big enough for harvesting.

RIPE. Okra pods are ripe at 2 to 3" in length.

OPTIMAL PLANTING AND HARVEST TIMES

Okra, from Transplants									Suitable for Succession Planting?		N

Weeks Before Last Spring Frost			Last Spring Frost				Weeks After Last Spring Frost						
-6	-4	-2	0	+2	+4	+6	+8	+10	+12	+14	+16	+18
				🌱			🧺						

🌱 = Planting Period   🧺 = Harvest Period

ONIONS

YOU SHOULD HAVE NO PROBLEM MAKING good use of all the onions you can grow, given their value in the kitchen. Red, white, and yellow onions are all harvested in the same way. All onions like hot temperatures and need to come out of the ground before cooler, wetter fall weather sets in.

Ripe is really a matter of how you want to use the onions. If you need scallions, pull up immature onions as soon as the greens are big enough. The flavor will be mild.

Let the onions mature by waiting until the tops yellow and fall over, and the onion shoulders begin to push up through the soil. At that point, use a rake to break the tops and bend them over, which will help speed ripening.

Harvest the onions on a sunny morning, after the tops are completely brown. Carefully loosen the soil (even a minor bruise can turn into rot in a fresh onion) and then pull the onions up. Leave them to dry in the sun for the rest of the day.

If an onion sends up a flower stalk, harvest the onion immediately and use it as soon as possible.

UNDER-RIPE. Unripe onions can be used as scallions as soon as the greens are big enough.

OPTIMAL PLANTING AND HARVEST TIMES

Onions, from Seeds								Suitable for Succession Planting?			N
Weeks Before Last Spring Frost			Last Spring Frost				Weeks After Last Spring Frost				
-6	-4	-2	0	+2	+4	+6	+8	+10	+12	+14	+16	+18

Harvest time depends on use. New "scallion" onions can be harvested within a few weeks of last spring frost. For full bulb onions, wait until fall.

⌒ = Planting Period   ⎍ = Harvest Period

IN THE MARKET

It's easy to determine ripeness in market onions.

Feel an onion by holding it like a baseball. It should be hard all over. Even small soft spots are indicators of possible areas of rot underneath. The outer skin should be intact but dry and brittle.

Smell the onion. Uncut fresh onions have a very subtle, almost non-existent odor. An overripe onion, on the other hand, will have a strong, overpowering odor.

Look for any cuts or blemishes on the surface—signs of potential rot. Healthy fresh onions will have a stem that is tightly closed and dry.

EXTENDING RIPENESS

Once dry, brush the soil off harvested onions and trim the tops to about 1 inch. Cure the onions on screens in a cool, dark, dry, well-ventilated area such as a garage with operable windows or vents. Cure them for two to three weeks, until the skins become brittle. Store the onions in a cool, dry area, in temperatures between 40°F and 50°F. Onions will keep for several months when dried and stored properly.

RIPE. Let the onions mature by waiting until the tops yellow and fall over.

OREGANO

THIS WONDERFUL PERENNIAL HERB IS essential to Italian, Greek, and other Mediterranean cuisines, and is used fresh and dried. Harvest leaves by removing just a few from different areas, or cut off whole stems down to just above a leaf pair. Harvest any time the plant is more than 4 inches tall.

Cut leaves midmorning when the oil concentration (and flavor) will be at its highest. The flavor of the leaves becomes more intense the closer the plant gets to flowering. The flowers have a subtler flavor and are wonderful on salads.

IN THE MARKET

Look for fresh oregano with bright, perky leaves. Avoid any with browning leaves, mold, or containers with leaves that have fallen off the stems. For the strongest flavor, buy dried oregano.

EXTENDING RIPENESS

Use oregano fresh when picked, or dry it for more intense flavor. Tie several stems into a bundle and hang them in a cool, dark, well-ventilated location. Strip the leaves after they're dried. Crush the dried leaves and store in airtight plastic or glass containers.

UNDER-RIPE. Oregano is ripe as soon as plants are 4" tall. Harvest oregano by removing just a few leaves from different areas.

OVERRIPE. Although this plant is past its prime, the flowers have a subtle flavor that is wonderful on salads.

OPTIMAL PLANTING AND HARVEST TIMES		
Oregano	**Suitable for Succession Planting?**	**N**
Not applicable. Oregano is a perennial herb that can be harvested almost anytime..		

PARSLEY

FRESH CURLY PARSLEY HAS A MILD, SOMEWHAT grassy flavor. Flat-leaf Italian parsley has a more complex flavor. Both are cultivated and harvested in the same way.

Regularly trimming plants spurs growth. Start harvesting when the plant reaches 6 inches tall. Cut individual stems at the base of the plant. Harvest from the outside in, because new growth sprouts from the center.

IN THE MARKET

Both types of parsley are usually sold in bunches. The leaves should be crisp and perky, evenly green all over, and there should be no sogginess or sliminess inside the bunch. Make sure there is no browning on the cut ends of the stems, which would indicate a long time between harvest and market.

EXTENDING RIPENESS

Fresh-cut parsley should be used as soon as possible, although you can keep stems or bunches for several days in a glass of cool water. You can also refrigerate it by rinsing, shaking off excess water, and putting it in a vegetable crisper in a loose plastic bag. It will last for up to one week.

RIPE. Flat-leaf parsley has a complex flavor. Regularly trimming parsley plants spurs their growth.

RIPE. Curly parsley has a mild, somewhat grassy flavor.

OPTIMAL PLANTING AND HARVEST TIMES

Parsley, from Seeds								Suitable for Succession Planting?					N
Weeks Before Last Spring Frost			Last Spring Frost	Weeks After Last Spring Frost									
-6	-4	-2	0	+2	+4	+6	+8	+10	+12	+14	+16	+18

🌱 = Planting Period 🧺 = Harvest Period

SPOILER ALERT:
7 FRUITS AND VEGETABLES WITH THE SHORTEST LONGEVITY

Some fruits and vegetables simply don't have staying power. They're meant to be enjoyed in the moment because their ripeness passes so quickly.

1. Berries.

Although strawberries are somewhat more durable, even they are subject to quick spoilage. The shortest shelf life belongs to raspberries. They can turn to mush in two days. Blueberries and blackberries are only slightly more durable.

2. Bananas.

Spoiling isn't really an accurate term, because even when they turn to mush, brown bananas can still be kitchen gold. They are great at ripening themselves, and don't slow down even after they are completely brown.

3. Avocados.

The fruit ripens quickly off the tree, and it can seem like, given the tough protective shell, they're immune from further softening, but they can rot within days, especially if bruised.

4. Asparagus.

There is a lot of space on an asparagus spear through which moisture can evaporate and leave the spear limp and unappealing. That's why cut asparagus spears fade quickly.

6. Watercress.

As the name suggests, the plant needs hydration. Leaves and stems removed from a source of water quickly wither and go unappealingly limp. Watercress rarely lasts more than two days after harvest or purchase.

5. Corn.

Even inside the husk, an ear of corn is busy deteriorating from the moment you pick it. The corn rapidly converts delicious sugar in the kernels to tasteless, glumpy starch. It will last a week or more refrigerated, but will have lost most of its flavor and crisp, fresh texture.

7. Basil.

Fragile basil leaves surrender wonderful flavor when cut or chewed. But that fragility is a weakness. The leaves quickly lose vigor once removed from the plant, which is why it's wise to grow a windowsill pot.

PEAS

PEAS ARE POPULAR AMONG HOME GARDENERS, who choose from three basic types.

Sweet peas, also known as English or shelling peas, are picked and then shelled because the pod is too stringy and tough to eat. Eat as soon after picking as possible, because sugar in the peas converts rapidly to starch once harvested. Pods are ripe when full and rounded, but shouldn't be overly large. You should be able to feel the plump, fully formed peas inside. If in doubt, pick the plumpest pod and break it open. If the peas inside are full and firm, use that pod to judge the others for ripeness.

Sugar snap peas grow in edible pods. The pod should be plump and well formed, but not too large. Overripe pods are tough and stringy. The pod is ready for harvest if it audibly snaps when you break it.

Snow peas are a cross between snap and English peas. They should be picked young, when they are exceedingly tender with a light, refreshing flavor. The pod will be thin and bright green, and the peas small and not well filled out.

To pick pea pods, hold the vine with one hand and pinch the pod off where it connects with the stem, using your thumb and forefinger. Pick in the morning.

IN THE MARKET

Buy fresh shelling peas only at a farmers' market or a farm stand. Look for pods that are full and plump, but not overly large. They should be evenly green with no brown or yellow spots.

Snap peas should be firm and evenly colored, and should snap rather than bend. Snow peas should be light, bright green with no discoloration or scars. The pod should have very little give. Avoid limp snow peas.

EXTENDING RIPENESS

Sweet peas are best shelled and cooked the day they are picked. To preserve them, shell and blanch the peas, then freeze them in a tightly sealed container.

Snap peas will keep for several days refrigerated in a sealed plastic bag. Snow peas will keep for up to two days. But both are best used as on the day they're harvested or bought.

OPTIMAL PLANTING AND HARVEST TIMES

Peas, from Seeds								Suitable for Succession Planting?		Y
Weeks Before Last Spring Frost			Last Spring Frost	Weeks After Last Spring Frost						
-6	-4	-2	0	+2	+4	+6	+8	+10	+12	+14	+16	+18

= Planting Period   = Harvest Period

RIPE. Sweet peas should be eaten as soon as possible after picking. Sweet peas are picked then shelled because the pod is too stringy and tough to eat.

RIPE. Snap peas should be firm and evenly colored.

RIPE. Sugar snap peas should be plump and well formed.

RIPE. Ripe snow peas are thin and bright green. Snow peas should be picked young.

PEPPERS

PEPPERS CAN BE PICKED WELL BEFORE maturity, based on the flavor, sweetness, and color you prefer.

Bell peppers range from green to deep purple depending on variety; when ripe, the pepper is the variety's intended color, and will be large and thick-walled. Sweetness and vitamin C are most pronounced in a mature pepper. Bell peppers are commonly picked green to speed harvest, spur new growth, and to suit personal preference or recipes.

Hot peppers are hottest at maturity, but they, too, can be picked early to increase yield and moderate heat and flavor. Common types include:

- Jalepeños are usually picked slightly young, because maturing doesn't noticeably change heat and flavor. Harvest jalepeños from 2 to 3 inches long, mid- to dark green, and firm with smooth skin.

UNDER-RIPE. The nascent pepper will appear almost immediately after the flower falls away. From this stage, ripe fruit is only two or three weeks away.

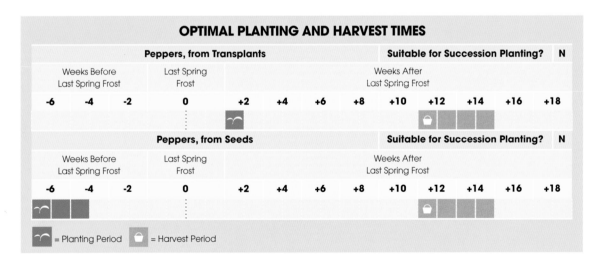

OPTIMAL PLANTING AND HARVEST TIMES

Peppers, from Transplants													Suitable for Succession Planting?	N
Weeks Before Last Spring Frost			Last Spring Frost	Weeks After Last Spring Frost										
-6	-4	-2	0	+2	+4	+6	+8	+10	+12	+14	+16	+18		
				⌣					🪣					

Peppers, from Seeds													Suitable for Succession Planting?	N
Weeks Before Last Spring Frost			Last Spring Frost	Weeks After Last Spring Frost										
-6	-4	-2	0	+2	+4	+6	+8	+10	+12	+14	+16	+18		
⌣									🪣					

⌣ = Planting Period 🪣 = Harvest Period

RIPE. Red bell peppers (bottom) at ripeness are large and thick-walled. One bell pepper plant can have peppers at varying stages of ripeness.

- Chili or cayenne peppers grow up to 8 inches long. Personal preference for heat level will determine when you pick them. Color signals maturity, as the peppers grow increasingly red from tip to stem. If you're drying the peppers, pick them totally red but still firm.

- Habaneros are classic hot peppers. A ripe habanero is round, from 1 to 2 inches in diameter, and either a beautiful orange or red. Because habaneros are grown for their heat, they are picked mature. If minute brown lines form on the pepper, it has peaked and should be picked immediately. Picking slightly early increases yield. A ripe habanero will come off the plant with a gentle tug.

- Poblanos are hard to judge for ripeness because the mature color is just a deeper shade of the immature green. Go by size. A ripe poblano should be about 4 inches long and 2 inches across. Pick them mature for a more complex flavor and restrained heat.

- Serrano peppers, when ripe, are five times hotter than jalapeños. Many cooks prefer to pick serranos when green and slightly small because the heat is less intense. If you're fond of the burn, pick the pepper when bright red all over.

RIPE. Jalapeños are harvested at 2 to 3" long. Wear latex gloves when harvesting ripe hot peppers.

RIPE. Yellow bell peppers are picked once they turn bright yellow. They can be easily twisted off the plant.

Ripe bell peppers can easily be twisted off the plant, but it's best to use a sharp knife or garden shears to cut the stem, leaving 1 inch of stem attached to the pepper.

Cut hot peppers off the stem to avoid damaging the plant. Wear latex gloves when harvesting ripe hot peppers because oil from the peppers can transfer from your hands and irritate the eyes, nose, and mouth.

IN THE MARKET

Peppers of any kind are easy to damage during harvest. Inspect market peppers for any nicks or mars on the surface. The pepper should be shiny and the color lively. Avoid any with soft spots.

EXTENDING RIPENESS

Bell peppers stay fresh for a week refrigerated in a loose plastic bag. Hot peppers vary on how long they'll last refrigerated. But to enjoy the best heat and flavor from hot peppers, use them within two days of harvest.

Dry hot peppers to concentrate heat and flavor and add a slightly smoky note. Dry them in a dehydrator, or thread a string through the peppers and hang them in a shaded, warm, dry room for two to three weeks. Grind or chop dried peppers and use them as spices, or leave whole and reconstitute them by soaking in cool water for about 30 minutes.

POTATOES

HARVESTING POTATOES IS EASY, BUT RIPENESS is subject to personal interpretation. Some gardeners prefer sweeter "new potatoes" that are dug up early, before the tubers' sugars have finished converting to starches. You can dig up new potatoes once the plants start flowering.

But to harvest truly ripe potatoes, wait until the vines begin to die. As the tops die back, dig up a test potato. The skin should be firmly attached; if it can easily be rubbed off, the potatoes are not yet mature.

Ideally, dig up potatoes using just your hands to avoid damaging the potatoes. If the soil is too firm, use a pitchfork to loosen it from the edges of the bed, being careful not to damage the tubers. Harvest on a dry day to make the process easier.

When all the vines die, or if a hard frost is due, harvest all remaining potatoes to prevent rotting or spoilage.

Cure potatoes by first laying them out in a warm, partially shaded area with good airflow, for at least 30 minutes on each side. Once dry, any dirt should be easy to brush off. Use the potatoes as soon as they're cured, and use cut or damaged potatoes immediately. Never wash potatoes until you're ready to use them, because washing will shorten the effective life of the potato in storage.

UNDER-RIPE. Young potato plant.

OPTIMAL PLANTING AND HARVEST TIMES

Potatoes	Suitable for Succession Planting?	N
Not applicable. Potatoes are grown from seed potatoes planted in early to late spring and harvested all at once.		

THE BEST POTATO

Different types of potatoes have different concentrations of starch, which affects flavor and texture and determines the best cooking method. For instance, starchy potatoes, such as the classic Russet, are typically mashed or made into French fries and won't have an appealing texture if cut and roasted. The chart below provides a shopping guide based on preferred preparation.

Potato	Starch Content	Best uses
Idaho/Russets	High	Baking, frying, mashing
Red	Med to low	Roasting, salads
Yellow	Medium	All-purpose
Yukon Gold	Med to low	Steaming, boiling
Purple	Medium	Steaming, boiling
Fingerlings	Low	Roasting, salads
White	Medium	All-purpose

IN THE MARKET

Look for potatoes that haven't been completely cleaned and those with uniform shapes that will be easy to peel or chop. Avoid any with green areas or sprouts. Pass over potatoes with wrinkled skins, soft spots, bruises, visible cuts or damage, or any decay. The potato should feel slightly heavy for its size.

EXTENDING RIPENESS

Never store potatoes with apples, tomatoes, or bananas (the ethylene gas emitted by those fruits can cause the potatoes to spoil). Store potatoes in a cool, dark place, such as a paper bag in the pantry or a cupboard. Don't store them in plastic or in the refrigerator.

For long-term storage, use a root cellar or a similar environment. Store them, after curing, in a crate or other breathable container. Keep them out of the light, at a temperature between 40°F and 45°F. Stack them no higher than three layers, to avoid bruising or other damage. The potatoes should keep until the following spring, when they'll naturally begin to sprout.

> **RIPE.** If the skin can be easily rubbed off, the potatoes are not yet mature.

PUMPKIN

HELP PUMPKINS RIPEN BY REMOVING LEAVES that shade them as the leaves turn yellow and brown. To judge ripeness, it's important to know the mature size and color for the variety you've planted.

- **Color.** Traditional golden orange is most common, but white, light orange, and even green pumpkins are available.
- **Size.** Mature size can be affected by a lack of soil nutrients, underwatering, and other factors.
- **Rind hardness.** A ripe pumpkin's rind is hard enough to resist a fingernail from puncturing the surface.
- **Stem.** The stem of a pumpkin nearing maturity will begin to shrivel and dry out.

Harvest pumpkins by cutting them from the vine with a sharp knife or loppers. Wear long sleeves and gloves, because pumpkin vines can be prickly. Leave at least 3 inches of stem on the pumpkin to protect against infection. Handle harvested pumpkins carefully because they bruise easily.

IN THE MARKET

The ideal pumpkins have a pleasing uniform shape rather than an unbalanced look. The color should be rich and bright, and the surface of the pumpkin should be free from cuts, cracks, scrapes, and other imperfections.

∧ **UNDER-RIPE.** Small unripe fruit on a pumpkin plant.

> **RIPE.** A ripe pumpkin is commonly orange with a thick rind. The stem shrivels and dries out as the pumpkin nears maturity.

OPTIMAL PLANTING AND HARVEST TIMES													
Pumpkin (Winter Squash), from Transplants								Suitable for Succession Planting?					N
Weeks Before Last Spring Frost			Last Spring Frost	Weeks After Last Spring Frost									
-6	-4	-2	0	+2	+4	+6	+8	+10	+12	+14	+16	+18	
				〰				🛍					

〰 = Planting Period   🛍 = Harvest Period

EXTENDING RIPENESS

If you're using pumpkin fresh, simply wash off the rind and keep it on the counter for a day or two until you're ready to cook the flesh. You can save the seeds to use for the next year's crop, or to eat. In either case, scoop them out and rinse thoroughly in a colander. If you're saving the seeds for planting, spread on a flat baking sheet and turn every day for about a month. Then store in a cool, dark area inside an airtight container.

If you're eating the seeds, spread them in a single layer on a greased baking sheet. Roast them in a 300°F oven for about 30 minutes, and then toss with salt and whatever spices you prefer. Dried pumpkin seeds will last for years.

To store pumpkins long term, cure them in a sunny spot for 10 to 14 days until the skin has hardened. Store them in a cool, dry area, such as a root cellar, on a slatted platform, rather than a floor. They shouldn't touch in storage. The ideal temperature is around 55°F. Never store pumpkins with apples because the ethylene gas produced by the apples will decrease the pumpkins' lifespans. Immediately remove any pumpkins that show signs of rotting.

RIPE. A ripe delicata squash is cream-colored with green stripes.

THE LESS-FAMOUS WINTER SQUASHES

Pumpkin isn't the only winter squash. Popular and widely available winter squashes include:

Squash	Description	Harvest & Storage
Butternut	Bell-shaped squash that offers silky flesh, almost candy sweet with a mild nutty flavor.	Harvest when the rind is hard and deep tan, in the same way you would pumpkin. Cure butternut two weeks at 70°F indoors. It can be stored up to six months at around 45°F.
Acorn	An unusual squash with a thicker rind, acorn squash looks like its namesake.	Ripe when a dark green color—ground color will change from yellow to orange. The stem withers and browns. Harvest late rather than early, in the same way you would pumpkin. Don't cure; it will keep for several months in a dry, well-ventilated area at around 50°F.
Hubbard	Sometimes called "green pumpkin" for its sickly hue when ripe, hubbard has a flavor similar to pumpkin.	When ripe, the first few inches of stem will dry and look like cork. If the vine begins dying, the squash is ready. Hubbard is harvested in the same way as pumpkin, and is cured for two weeks before storing at around 50°F.
Spaghetti squash	Yellow, barrel-shaped squash with flesh that cooks up to resemble pasta.	Ripe skin color changes from cream white to pale yellow; a ripe rind will harden so much it resists scratching. Harvest as you would pumpkin. Cure at 70°F for one week. Store long term in a cellar or garage at a consistent 50°F.
Delicata	Beautiful small squash, cream with green stripes when ripe. Has tender skin and tastes like sweet potato.	Harvest when the skin hardens so that it won't dent under your fingernail and the stem dies completely. Harvest in the same way as pumpkin, and store at 50°F for up to several months.

RADISHES

A COOL-SEASON FAVORITE, PERFECT FOR A child's garden, radishes are some of the fastest-growing garden crops. Some spring varieties can be ready for harvest in as little as three weeks. Winter radishes take longer to mature and tolerate being in the ground longer. Spring radishes should be picked early rather than late, because they become spongy and unpalatable when left in the ground too long. Winter varieties are favored by most gardeners because they keep longer in the refrigerator and have more complex flavors. No matter what type you're planting, the longer the radish stays in the ground, the spicier it becomes.

Check your radishes frequently to determine if they're ready for harvest (daily is best). Remove a little soil around the top and check the diameter. Most varieties should be about 1½ inches in diameter—although go by the recommended mature size on the seed packet for the variety you've planted.

Harvesting could hardly be easier. Grab the base of the greens and slowly and steadily pull the radish up out of the soil.

IN THE MARKET

The greens are a key indicator of freshness in bunches of produce aisle radishes. If they are perky and bright, the radishes are fresh. Avoid any bunches with greens that are wilted, yellowed, or turning mushy or brown.

Feel the radishes. They should be rock hard.

Look for insect damage, apparent in neat, small holes, or other unusual blemishes in the surface of the radish.

OVERRIPE. Left a little too long, the aboveground shoulders will split and the radish will become unpleasantly spicy.

OPTIMAL PLANTING AND HARVEST TIMES

Radishes, from Seeds								Suitable for Succession Planting?					Y
Weeks Before Last Spring Frost			Last Spring Frost					Weeks After Last Spring Frost					
-6	-4	-2	0	+2	+4	+6	+8	+10	+12	+14	+16	+18

⌣ = Planting Period   ⌂ = Harvest Period

RIPE. To harvest a ripe radish, grab the base of the greens and slowly and steadily pull up.

EXTENDING RIPENESS

Because one of the key benefits of a radish is its crispness, it's best to use them the day they are harvested or purchased. However, they will store in a loose plastic bag in the refrigerator for up to 10 days. Cut off the greens and wash the radishes before storing. You can reconstitute limp, rubbery radishes by soaking them in ice water for 15 minutes or more.

Don't discard the greens. They have a nice, light peppery flavor that makes them a wonderful addition to salads and salsas. Radish greens will last up to two days in the refrigerator, but it's best to use them immediately.

SAGE

LET SAGE ESTABLISH IN THE FIRST YEAR, with just light harvesting, and you can harvest as the need arises the second year and beyond. Don't harvest during the winter.

Pick sage leaves when there has been no rain, in the morning after the dew has dried. Pinch individual leaves as needed or cut an entire stem. To keep the plant vigorous, trim 6 to 8 inches off the top of the plant at least twice in a season.

IN THE MARKET

Sage sprigs are sold in clamshell packaging. Pick out sprigs with leaves that are still full and lively, not curled or wilting. Open the package and the sage odor should be strong. Avoid any with cut leaves, or white, brown, or black spots.

EXTENDING RIPENESS

Sage is best fresh, but you can dry it in a warm, dark, well-ventilated area, such as a garage or shed. Spread individual leaves on a screen, or bundle and tie sprigs and hang them. You can also freeze the leaves, which many cooks believe better preserves the flavor in the leaves.

RIPE. To harvest sage, pinch individual leaves as needed or cut an entire stem.

OPTIMAL PLANTING AND HARVEST TIMES

Sage, from Seeds				Suitable for Succession Planting?	N

Weeks Before Last Spring Frost			Last Spring Frost	Weeks After Last Spring Frost									
-6	-4	-2	0	+2	+4	+6	+8	+10	+12	+14	+16	+18

= Planting Period   = Harvest Period

SCALLIONS

MOST ONIONS CAN BE PICKED IMMATURE AS scallions, but *Allium fistulosum* is the species grown specifically as scallions. The longer it matures in the ground, the stronger the flavor, but wait too long and you'll have an onion. Start harvesting as soon as there is about 6 inches of top growth.

Harvesting is easy thanks to the slender shape. Firmly grasp the scallion about an inch up from the soil and pull it out of the ground with even, steady pressure. If the soil is sticky, loosen it gently with a small garden fork, being careful not to damage the delicate scallions.

IN THE MARKET

Scallions are sold in small bunches; inspect each scallion closely. The white part should be bright and the roots should still be attached. Avoid any that are limp or nicked, and any with limp, yellowed, or brown top growth.

EXTENDING RIPENESS

Don't wash scallions until you're ready to use them. Store in a sealed plastic bag in the refrigerator's vegetable drawer. They should keep for four to seven days.

RIPE. Scallions are a form of onion intended for early harvest. If you leave scallions in the ground too long, you'll have an onion.

OPTIMAL PLANTING AND HARVEST TIMES		
Scallions	**Suitable for Succession Planting?**	**Y**
Too variable to be reliably charted.		

SPINACH

THIS COOL-SEASON GREEN IS A PROLIFIC producer and it's easy to harvest. Individual leaves are ripe as soon as they are large enough to use in the kitchen. Cut off outer leaves with scissors by snipping stems where they connect to the plant. But plants must be watched carefully in order to harvest at their prime. If the weather turns hot and the plant threatens to bolt, cut the entire plant off at the base. Harvest fall or winter varieties until the first hard frost.

IN THE MARKET

Increasingly, fresh spinach is sold in bunches with roots attached, which helps leaves retain moisture. Pre-packaged spinach has a shorter lifespan than loose leaves. In either case, the leaves should be deep, vibrant green, not yellow, limp, slimy, or damaged.

EXTENDING RIPENESS

Spinach shouldn't be washed before being stored. The rich nutrients peak when the leaves are freshest, so eat spinach as soon as possible after harvest.

You can refrigerate spinach in a loose plastic bag for several days. Wash spinach right before using it. Rinse thoroughly in a large bowl of water until the water comes out clean, and then spin dry in a lettuce spinner.

ʌ **OVERRIPE.** At the point where spinach plants begin to send flower shoots, they are past prime.

> **RIPE.** Individual spinach leaves are ripe as soon as they are large enough to use in the kitchen.

OPTIMAL PLANTING AND HARVEST TIMES

Spinach, from Seeds								Suitable for Succession Planting?				Y

Weeks Before Last Spring Frost			Last Spring Frost	Weeks After Last Spring Frost								
-6	-4	-2	0	+2	+4	+6	+8	+10	+12	+14	+16	+18

 = Planting Period   = Harvest Period

STRAWBERRIES

STRAWBERRIES DON'T RIPEN AFTER YOU pick them. In fact, they deteriorate in quality from the moment the strawberry is picked as the natural sugars begin to turn into flavorless starch. Ripening and harvesting doesn't differ between day-neutral, everbearing, and June-bearing varieties. Ripe berries will be uniformly red, with no whiteness or green spots.

Harvest strawberries in the morning. Look for fruit that is completely red. You can harvest by snipping the ripe berry stem with sharp scissors. However, experienced gardeners pick the fruit by hand, by cupping the berry loosely in the palm of the hand, holding the stem slightly up from the berry between the forefinger and thumbnail, and gently pulling and twisting at the same time. Some strawberry varieties are bred to "cap" easily. If you are growing one of these, hold the berry and apply pressure right behind the cap. A ripe strawberry will easily pull away, leaving the cap intact on the stem.

Keep newly harvested strawberries out of the sun, and don't crowd the harvesting container, because strawberries bruise easily.

Remove and discard damaged or misshapen berries as you harvest the ripe strawberries to prevent any rot or disease from spreading.

RIPE. You can harvest strawberries by cutting the stem with a sharp pair of scissors; experienced gardeners pick the fruit by hand.

OPTIMAL PLANTING AND HARVEST TIMES		
Strawberries	**Suitable for Succession Planting?**	**N**
Not applicable; strawberries are grown from mature plants, which can take up to three years to fruit. Harvest time in most regions is April through June.		

ALL ABOUT ALPINES

Alpine strawberries are the smaller, more delicate cousins to the traditional strawberry, with more intense flavor including pineapple and floral tones. They have a light and elegant fragrance. The downside? Small harvests. You'll harvest less than a small handful at any one time. Alpine plants produce a continuous harvest. The fruit is harvested in the same way as other strawberries are. Ripe alpine strawberries are deep red and fragrant. Use the tiny fruits immediately, because they hold up poorly in any kind of storage.

IN THE MARKET

The time a supply chain takes means supermarket strawberries are rarely going to taste as sweet or flavorful as those you'll find at a farmers' market or roadside stand.

Aroma is not relevant, because only certain types are fragrant.

Size is not an indicator of quality. Big or little, ripe strawberries will taste just as wonderful and sweet. Choose a size that suits what you're preparing. For instance, extra large are best for chocolate-covered strawberries, where smaller types are good for strawberry shortcake.

Appearance speaks volumes. Green areas are signs that the berry was picked before peak ripeness and it won't be fully sweet. Stains on the packing cartons indicate overripe berries that have been crushed or are rotting. Look closely at the surface sheen and color. Strawberries past their prime will be darker and duller.

EXTENDING RIPENESS

If you want the sweetest strawberries, eat them the day you buy or harvest them. That's less of a concern if you're baking or cooking with the strawberries, especially if you are adding sugar to a dish like berry cobbler. You can store unwashed berries in the refrigerator, uncovered, for up to three days.

Strawberries also freeze well. Wash lightly, let dry, hull, and put them in a resealable plastic bag, squeezing out as much air as possible. They'll keep in the freezer for up to three months.

SWEET POTATOES

YOU CAN HARVEST SWEET POTATOES ANY time after they get big enough to eat, but it's best to wait until after the leaves have started to yellow. For the largest and most nutritious sweet potatoes, wait until the first fall frost is imminent. However, if cold weather turns the leaves black, harvest all the potatoes immediately.

To harvest a whole bed, cut away the vines at soil level. Then carefully loosen the soil around the perimeter of the bed using a garden fork—or the version known as a potato fork.

Dig out the potatoes with your hands and comb through the soil after you're done to check for any you missed. You can also harvest tubers from a single hill by cutting away the vines and locating the primary growing crown. Use that as a centerpoint and fork the soil in a 20-inch-diameter circle around the crown. Handle the potatoes gently, because they are susceptible to bruising before they're cured.

IN THE MARKET

Shop for sweet potatoes that are small to medium, because they'll be sweeter. Look for tubers with unblemished skins and even coloring. The skins should be smooth to the touch. A ripe sweet potato is heavy for its size, and firm all the way around. There should be no wrinkles on the skin and no sprouts.

EXTENDING RIPENESS

Freshly harvested sweet potatoes should be left to dry in direct sunlight for three to four hours. Shake off any dirt. You can use the potatoes immediately, but they will be sweeter and more nutritious if cured. Curing can heal damage like cuts, and uncured sweet potatoes don't bake properly.

Spread the sweet potatoes out on newspaper in a warm (around 85°F), dark area of the house, so that none are touching. Cure the potatoes for about 10 days.

Refrigerate store-bought sweet potatoes for up to two weeks. To store longer, wrap each sweet potato in newspaper and carefully pack in paper bags, baskets, or boxes. Store in a root cellar or similar conditions where the temperature does not vary much from 55°F. Sweet potatoes will last six months or more when stored under ideal conditions.

> RIPE. Harvest sweet potatoes any time after they get big enough to eat.

OPTIMAL PLANTING AND HARVEST TIMES		
Sweet Potatoes	**Suitable for Succession Planting?**	**N**
Not applicable. Sweet potatoes are grown from mature plants that can take up to three years to fruit.		

SWISS CHARD

PICK YOUNG SWISS CHARD LEAVES AS BABY greens for a salad. Younger leaves are more tender and flavorful. Harvest outer leaves continuously as they mature. You can cut them at full height—depending on the variety, from 1 to 2 feet tall. But tenderness and flavor will be just as good when the leaves are 8 to 12 inches tall.

To harvest, cut leaf stems 2 inches above the ground using a sharp knife. Cut only the outside leaves and you'll continue harvesting the plant until the first heavy frost. Never cut more than one third of the leaves.

IN THE MARKET

Look for swiss chard with glossy, wavy green leaves and bright, firm stems. Avoid any with spots or holes in the leaves. The cut marks at the end of the stems should look like they've just been made.

EXTENDING RIPENESS

Use swiss chard as soon as possible. Refrigerate unwashed, in a loose plastic bag, for up to a week. Right before using, wash the leaves well—dirt hides in the surface.

RIPE. Harvest swiss chard with glossy, wavy green leaves and bright, firm stems.

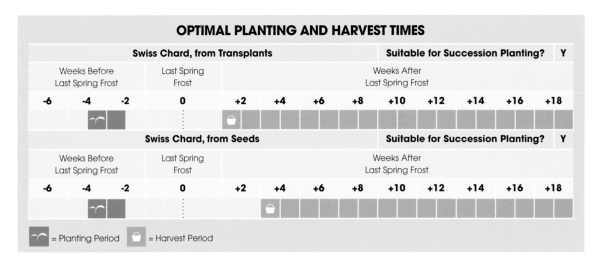

OPTIMAL PLANTING AND HARVEST TIMES

Swiss Chard, from Transplants													Suitable for Succession Planting?	Y
Weeks Before Last Spring Frost			Last Spring Frost	Weeks After Last Spring Frost										
-6	-4	-2	0	+2	+4	+6	+8	+10	+12	+14	+16	+18		

Swiss Chard, from Seeds													Suitable for Succession Planting?	Y
Weeks Before Last Spring Frost			Last Spring Frost	Weeks After Last Spring Frost										
-6	-4	-2	0	+2	+4	+6	+8	+10	+12	+14	+16	+18		

⌢ = Planting Period 🪣 = Harvest Period

THYME

THYME'S FLAVOR PEAKS RIGHT BEFORE THE plant flowers, but you'll realize great flavor by harvesting anytime as the plant grows.

Cut sprigs and strip them of the leaves (young, tender sprig ends can be chopped with the leaves). Harvest in the morning, cutting the stems just before a growth node or pair of leaves with sharp scissors.

IN THE MARKET

Thyme is most commonly sold in plastic clamshell packaging. Look for green stems and fresh, perky leaves. Avoid packages in which the leaves are browning, or where excess moisture is trapped.

EXTENDING RIPENESS

To use thyme immediately, rinse sprigs and let them dry naturally. Refrigerate unwashed thyme sprigs in a loose plastic bag for up to a week.

Dry sprigs in a dehydrator or hang them, tied in bundles, in a warm, dark place with good air ventilation. They should be dry in about a week.

You can also strip sprigs and dry individual leaves spread on a baking sheet. They should be dry within two days. Store dried thyme in an airtight container in a cool, dry area such as a kitchen cabinet.

OVERRIPE. When overripe thyme flowers, the herb begins to lose flavor.

RIPE. Thyme's flavor peaks right before the plant flowers.

OPTIMAL PLANTING AND HARVEST TIMES		
Thyme	**Suitable for Succession Planting?**	**N**
Not applicable. Thyme is a perennial that can be harvested throughout the growing season.		

TOMATOES

SEED COMPANIES AND NURSERIES PROVIDE A "days to harvest" estimate for tomato ripeness, but that figure is imprecise. Timing often depends on the variety, local climate, unusual weather events, and other variables.

Tomatoes continue ripening off the vine, so it's safe to harvest early. But there is an incomparable satisfaction in plucking a juicy, deep red, mature tomato off the vine and eating it still warm from the sun.

Color is the most obvious indicator of ripeness. Although red is the traditional color, tomatoes come in many enticing hues, from light, golden-yellow to deep, tantalizing purple. Know what color your variety is supposed to be when ripe. Watch for color changes closely. There is a very subtle difference between the color of a nearly ripe and perfectly ripe tomato.

Smell is a subtler indicator. Unripe tomatoes have virtually no smell. Once the mature fruits begin releasing ethylene gas, the fruit takes on a rich, savory odor that gets stronger the closer the fruit is to full ripeness.

Touch is key. A ripe tomato will always feel heavier than it looks. The tomato should still be firm while yielding subtly to pressure from your fingers. Tomatoes ripen from the inside out, so by the time it's completely soft, the tomato is overripe.

Harvesting ripe tomatoes is simple. Grab the tomato gently and cleanly twist it so that the stem breaks right above the "calyx," the green jester's cap on top of the tomato.

UNDER-RIPE. This variety of Big Boy tomato is ripe when pink; this beige fruit is approaching ripeness.

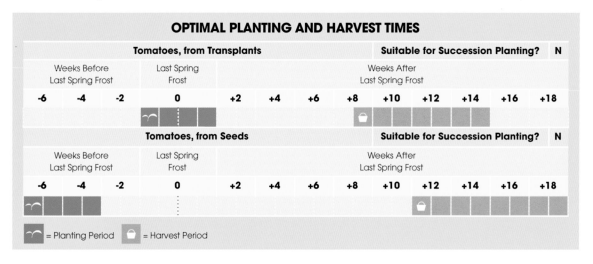

OPTIMAL PLANTING AND HARVEST TIMES

Tomatoes, from Transplants								Suitable for Succession Planting?				N
Weeks Before Last Spring Frost			Last Spring Frost	Weeks After Last Spring Frost								
-6	-4	-2	0	+2	+4	+6	+8	+10	+12	+14	+16	+18
			🌱				🧺					

Tomatoes, from Seeds								Suitable for Succession Planting?				N
Weeks Before Last Spring Frost			Last Spring Frost	Weeks After Last Spring Frost								
-6	-4	-2	0	+2	+4	+6	+8	+10	+12	+14	+16	+18
🌱									🧺			

🌱 = Planting Period   🧺 = Harvest Period

∧ RIPE. Picking a tomato while it is fully pink but not yet a deep red is perfect, as the fruit will come to full ripeness indoors and won't be subject to splitting or insect damage.

> OVERRIPE. Allowing a tomato to remain on the vine until it is deep red may cause the skin to begin to split.

HYBRID TOMATOES

These are the most common type of tomatoes, including such classics as Early Girl and Big Boy. Because the color is generally uniform, the most important indicators of ripeness will be smell and feel. Ripe hybrids have a distinctive, rich smell and feel firm with a slight give. Some large hybrids are meant to be harvested slightly early because truly ripe fruits may be so heavy as to damage the plant. This is particularly true of determinant varieties that—in contrast to indeterminate—ripen all their fruits in one short span.

HEIRLOOM

Heirloom tomatoes are open-pollinated varieties that have been passed down for at least fifty years through generations of gardeners. These feature the most unusual colors and color combinations of any tomato. The shapes are also irregular and unusual. Heirlooms get softer quicker than other tomatoes. Harvest them with "green shoulders," because they ripen unevenly, from the bottom up. Pick heirlooms when the bottoms are slightly soft to the touch and the tomato is heavy for its size.

CHERRY AND GRAPE

Ripe cherry tomatoes come off the vine with the slightest pressure. Err on the side of early picking, because if allowed to overripen, the tomatoes will split and draw fruit flies and other pests. When the first tomatoes in a cluster on a stem turn ripe, you can cut the entire stem. Hang it in the kitchen and pick the fruits one by one as they ripen.

IN THE MARKET

Look for evenly colored, firm—but not hard— tomatoes. Heirlooms will be softer. Avoid pre-bagged or packaged tomatoes because it's important to inspect each tomato, and the weight on the tomatoes in the bottom of a bag can cause bruising, soft spots, and even spoilage.

Check for soft spots. These indicate spoilage and a small soft spot can be a sign that much of the tomato will be unusable. For the same reason, avoid tomatoes at the bottom of a produce-bin pile.

EXTENDING RIPENESS

Frost is the enemy of all tomato plants. Some gardeners pull up the entire plant before the first frost, hanging the plant upside down in a cellar or a garage to allow the remaining tomatoes to ripen. However, it's easier to just pick unripe fruit and let the tomatoes ripen indoors.

- Store-picked tomatoes between 55°F and 75°F. Speed ripening by putting the tomato inside a paper bag or wrapping it in newspaper.
- Avoid direct sun. Sunlight doesn't aid ripening, but it does soften the tomatoes and can increase the chance of rotting.
- Don't let stored tomatoes touch or you'll increase the possibility of spot rotting.
- Do not refrigerate tomatoes—refrigeration destroys the texture and flavor of all tomatoes. (Temperatures below 55°F cause flavonoids to break down.)

RIPE. Ripe hybrids have a distinctive, rich smell and feel firm with a slight give.

RIPE. Heirloom tomatoes often look a little strange but are very tasty. These three tomatoes are at different stages of ripeness.

OVERRIPE. These cherry tomatoes, with skins that have split, have been left on the vine slightly too long.

RIPE. Grape tomatoes—some ripe, some under-ripe.

TURNIPS

HARVEST TURNIPS NOT ONLY FOR THE nutritious bulbs, but for the greens as well. The root is exceedingly tender and sweet when immature, but the flavor is more complex in mature turnips. The greens are even more nutritious and you can grow for that crop alone by crowding turnips and regularly cutting the greens after they grow to a few inches tall.

If you want to harvest both turnips and greens, take two or three leaves as the root grows and new leaves will grow. The more leaves you take, the more the growth of the root is slowed.

Harvest turnips between 2 to 3 inches in diameter. Larger bulbs can be woody and flavorless. Determine the diameter by brushing dirt back from the shoulders. You can simply pull up smaller turnips. For larger roots, loosen the soil with a garden fork, being careful not to touch the turnips. The turnips should easily pull right up. Harvest a fall crop before the first hard frost.

IN THE MARKET

Stores often sell turnips with the greens already removed, while farmers' markets sell them with the greens attached. It's usually best to buy them with the greens attached.

Look for bulbs no bigger than 3 inches in diameter. The turnips should be creamy white with light purple shoulders. There should be no nicks or cuts, discoloration, or soft spots. The bulbs should feel heavy for their size. If the greens are attached, they should be bright and perky, with no insect damage.

UNDER-RIPE. Young turnip plants don't yet have bulbs at their base.

OPTIMAL PLANTING AND HARVEST TIMES		
Turnips, from Seeds	**Suitable for Succession Planting?**	**Y**
Planting and harvesting are variable. Turnips are always grown from seeds outdoors, but can be harvested as greens, as young and tender roots, or as older roots.		

RIPE. Harvest turnips once they are between 2 and 3" in diameter.

EXTENDING RIPENESS

After harvesting turnips, cut off the greens. The bulbs will keep in a refrigerator crisper drawer for up to a month or store them in a root cellar for several months in crates lined with straw. (The turnips shouldn't be touching.) Keep them below 40°F, dry, and well ventilated.

Greens should be used within a couple days. Store them unwashed, in a plastic bag in the refrigerator. When ready to use, fill a large bowl or sink with cool water and swish the greens in the water. Grit will float to the bottom to be poured out. Refill with water and repeat until there is no grit in the water.

WATERMELON

A TRULY RIPE WATERMELON WILL HAVE THE typical symmetrical shape and large size, but the exact moment of peak ripeness is more about subtle signs.

Watch for the tendrils coming off the vine to turn brown, and for the "belly spot" (sometimes called "ground color"), where the watermelon rests on the ground, to turn from white to creamy yellow. (If you've used the SFG method of growing vertically on a trellis, you may not have this spot.)

Listen to the melon when you rap it. Immature watermelons make a pinging sound. Once ripe, it will be a softer hollow thump. You can get accustomed to the unripe sound by regularly rapping on the fruit as it grows. That will make discerning the ripe sound easier.

Once one watermelon is ripe, the others follow within a week or so. As they get close to ripe, cut back on watering to just enough so that you keep the vine leaves from wilting. This will concentrate the watermelons' sweetness.

To harvest a watermelon, cut the stem just above the melon end with a sharp knife. Handle harvested watermelons carefully to avoid bruising, which can cause the flesh to become mushy.

IN THE MARKET

Press on the surface of the watermelon with your fingertips. A ripe melon will have some give in the surface rather than be absolutely firm. However, the surface should also spring back against your fingertips, offering a bit of resistance. If it doesn't, the melon is overripe. You can also do the thump test, but you have to know what an unripe melon sounds like. You can also check that the ground color spot is rich yellow, rather than greenish white.

UNDER-RIPE. With fruit just budding, this watermelon plant has some way to go. Regardless, the appearance of tendrils and a thumping sound when rapped are stronger indicators of ripeness than size.

OPTIMAL PLANTING AND HARVEST TIMES

Watermelon, from Seeds								Suitable for Succession Planting?				N
Weeks Before Last Spring Frost			Last Spring Frost	Weeks After Last Spring Frost								
-6	-4	-2	0	+2	+4	+6	+8	+10	+12	+14	+16	+18

⌒ = Planting Period   🪣 = Harvest Period

RIPE. Rap on a watermelon to determine ripeness; ripe fruit will have a soft, hollow thump.

EXTENDING RIPENESS

Under optimal circumstances, a watermelon will keep up to three weeks. Store them in a cool, shaded area with little temperature variation (a temperature consistently around 65°F is ideal).

Do not refrigerate watermelons before cutting them. Wrap cut melons tightly in plastic and refrigerate. They should keep for five to seven days. You can also chunk and freeze watermelon for use in smoothies and sweet soups.

ZUCCHINI

ZUCCHINI PLANTS ARE PROLIFIC, SO YOU NEED to pay close attention to pick fruit at a tender, modest size and avoid tougher giant zucchini.

The leaves of the bushy plant grow to shade the fruits, so it's easy to miss a growing zucchini; the thumb-size fruit you remember seeing a few days ago is suddenly 12 inches long. Regularly move aside the leaves and thoroughly inspect into the center of each zucchini plant at least every two to three days.

The larger the fruit gets, the tougher the skin becomes and the more pronounced the seeds will be (the fruit is actually going to seed, in the same way that leafy greens bolt). The exceptions are cultivars that are specifically bred for size. If you're growing large zucchini on purpose, harvest them at the seed supplier's recommended mature size.

Once your zucchini grows to around 6 to 8 inches, cut the stem about 1 inch from the end of the zucchini, using a sharp knife. The plant will continue producing zucchini all season.

UNDER-RIPE. Tiny, young zucchini will surprise you with how fast they grow.

∧ **RIPE.** Harvest zucchini by cutting the stem about 1" from the end of the zucchini.

> **RIPE.** Harvest most zucchini at 6 to 8"; zucchini fruit become ripe in clusters.

OPTIMAL PLANTING AND HARVEST TIMES

Zucchini, from Transplants									Suitable for Succession Planting?			N
Weeks Before Last Spring Frost			Last Spring Frost			Weeks After Last Spring Frost						
-6	-4	-2	0	+2	+4	+6	+8	+10	+12	+14	+16	+18

〜 = Planting Period 　　 = Harvest Period

FLOWER POWER

Zucchini flowers are a delicacy. The plant produces both male and female flowers; harvest the males. (The females will have a small bud—a future zucchini—at its base.) Pick unblemished and perfectly formed flowers in the morning after they've opened, cutting them free and leaving 1 inch of stem. Leave some males on the plant. Harvest females if you'll grow more zucchini than you can use. (Females are tastier.) Use the flowers immediately, or seal them in a plastic or glass container with a moist paper towel for up to two days.

IN THE MARKET

Inspect zucchini in the store to find the best in the bin.

- **Size** is key. To use fresh or sauté, buy smaller zucchini. They'll be milder and sweeter, with a more delicate texture. Larger zucchini are good for zucchini bread or shredding.
- **Color** should be vibrant. Dull color is a sign of aging.
- **Damage**, such as nicks or cuts, is indicative of a zucchini to avoid. Some blemishes are to be expected, but any damage that exposes the flesh is a bad sign.
- **Check** for stems. A zucchini with a stem at least ½ inch long will last longer than one with no stem. Wrinkled skin is also a sign that the zucchini is past its prime.

Farmers' markets are the best places to find the freshest zucchini. Fresh-picked fruits will often be covered in bristly tiny hairs.

EXTENDING RIPENESS

Ripe zucchini have a high water content that translates to a very short shelf life. That's why it's always best to use your zucchini as soon as possible. Store ripe zucchini in the refrigerator, unwashed, in a loose plastic bag. They'll be good to use for about a week.

Although flavor and texture will suffer slightly, you can freeze zucchini for longer storage. Wash and slice or cube the zucchini, and then blanch it. Let it drain and dry, and freeze it in resealable plastic bags with all the air squeezed out. The zucchini will last for several months. The Food and Drug Administration recommends against canning zucchini because of the high potential for bacterial contamination.

OUTSIDE THE (SFG) BOX

ALTHOUGH MEL BARTHOLOMEW FELT SQUARE FOOT
Gardening is the best gardening method you could
choose, he'd be the first to admit that an SFG box has
its limits. For instance, the box doesn't accommodate
growing trees, and the SFG philosophy isn't really
about cultivating grow-and-forget perennials.

That means that your garden may include fruits and
vegetables that aren't ideal or won't fit in the SFG box.
That's okay. No matter where you grow and pick your
crops, you should still focus on the ripest possible fruits
and vegetables.

The pages that follow include listings for plants
that aren't naturally grown in an SFG box but are
commonly grown in gardens across the United States.

APPLES

PICKING APPLES AT THEIR RIPEST REALLY depends on how long you intend to store the apples. Start with the maturation date for your variety. Different apples mature in early, mid-, or late season, roughly correlating to midsummer, late summer, and fall. The later an apple ripens, the longer you can store it. But picking any apple far too early means an unpleasantly sour flavor and starchy texture. Overripe apples are mushy.

If you want to eat the apples immediately, pick those that are totally ripe. The apple will almost fall off into your hand. A slight twist is all you'll need to remove it from the branch. Healthy apples on the ground indicate others are ripe. Damaged and diseased apples may fall at any time (pick them up and discard them to prevent the spread of disease).

Feel the apple. If it's ready to eat, the apple will be firm, but with a slight softness or give. It's a good idea to feel your apples as they mature so you'll recognize the change in density that signals ripeness.

Check the ground color. Spots of green on a red or yellow apple are called the ground color, because it's the part that receives virtually no sun. On ripe red apples, the spot changes from green to creamy yellow or yellow green. On yellow apples, it changes from green to gold.

If you plan on storing the apples for a long period, pick them when they are the mature color but while the apple is still hard. Remove the apple with the stem intact.

Whether you're picking apples perfectly ripe or slightly before, handle them gently. They bruise easily, and soft spots and bruises can lead to rotting. (Rotting can pass from one apple to another in storage.)

IN THE MARKET

The best ripe apples are available midsummer through late fall. Large supermarkets sell apples that have been treated with a special gas and put in cold storage (it's why you can enjoy apples year-round). Although the quality is not noticeably deteriorated, these apples may be older than 10 months in some cases.

Inspect apples for damage. Obvious damage can indicate large areas of rot under the skin. Watch for small holes (insects); large browned or "burned" areas indicating sunburn; bruises from mishandling; or cuts that have penetrated the skin.

Color is a lesser indicator of ripeness. A ground color spot is okay, but otherwise the apple should be the variety's mature color.

Texture is tricky. A ripe apple is just slightly softer than an immature fruit. But an overripe apple will be noticeably soft.

Avoid apples that have sat at the bottom of the bin under layers of other apples. Stacking increases the risk of bruising and subsequent rotting. Reject apples with bruise spots.

OPTIMAL PLANTING AND HARVEST TIMES
Apples
Apple trees can take two to four years or more to mature and produce fruit. Harvest time will depend on variety and climate.

RIPE. Ripe apples will come off the tree with a slight twist. Harvest apples gently—they bruise easily.

EXTENDING RIPENESS

Whether you're eating or storing the fruit, use the largest apples first because smaller apples last longer. Do not store damaged apples. Cut out the damaged portion and use the remaining part as soon as possible.

Ripe apples last several days on a kitchen counter, but can last several weeks in the right conditions in the refrigerator. Store them near the bottom of the refrigerator, in a large, loose plastic bag—away from other fruits and vegetables. Prepared carefully, apples can be stored to last almost a year, but certainly six months or more.

The first step is to weed out any "bad apples." Damaged apples emit more ethylene gas, potentially spoiling the other apples. Pack the apples carefully. Wrap each apple in a sheet of newspaper, or layer them surrounded by straw or shredded paper in boxes or crates. (The apples shouldn't touch each other.) The cardboard trays used by commercial growers to transport apples in boxes are excellent for long-term apple storage.

Label boxes or crates with the date stored and the variety. Store the apples in a cellar or basement, around 40°F, with high humidity and good air circulation. Don't store apples near other produce, especially potatoes. Potatoes emit a gas that will cause the apples to ripen quickly and spoil in storage.

APRICOTS

APRICOTS CONTINUE TO RIPEN AFTER PICKING. So you can pick them ripe or harvest slightly "green" apricots.

Color indicates ripeness. The fruit changes from green to a dusty golden-yellow.

∧ **UNDER-RIPE.** Ripe apricots change from green to a dusty golden yellow.

> **RIPE.** Pick apricots by hand—a ripe apricot will come off the tree with a light twist; ripe apricots feel soft but not mushy.

Texture is a key "tell" to the prime picking moment. Ripe apricots feel soft but not mushy—exactly midpoint between rock-hard and nearly fall-apart mushy. Never squeeze apricots too hard because they bruise easily.

Smell the apricots. Ripe apricots smell distinctively sweet and floral.

Pick apricots by hand. A ripe apricot will come off the tree with a light twist. Immature apricots will be more resistant.

IN THE MARKET

You'll find apricots in season, late spring to late summer. They should be unblemished, light golden, and soft but not mushy. You can buy immature apricots and ripen them at home, but avoid any that are shriveled or small and rock hard.

EXTENDING RIPENESS

Ripe apricots last four to five days on the counter. Refrigerate apricots in a sealed container. They will keep up to two weeks.

Ripen immature apricots in a paper bag. They should be ready to eat in two to three days. Don't store them with bananas or apples.

OPTIMAL PLANTING AND HARVEST TIMES
Apricots
Apricot trees can take three to four years or more to mature and produce fruit. Harvest time will depend on variety and climate.

ARTICHOKES

ARTICHOKES ARE A RATHER FINICKY VARIETY of thistle. They grow on tall, sturdy stems. The central bud atop the stem may be followed by smaller artichokes growing off the main stem.

The trick is to let the bud grow as large as possible without opening. Pay close attention to the maturing bud, and inspect it daily as it nears maturity. Ripe, it will be 3 inches in diameter. You can harvest it smaller without losing flavor or tenderness.

Harvest a ripe artichoke with a sharp knife, cutting the stem cleanly 3 inches below the artichoke's base.

IN THE MARKET

Buying ripe artichokes is a matter of rejecting those past their prime. Avoid any that appear to be opening or look and feel dried out. Fresh, ripe artichokes squeak when squeezed and will feel heavy for their size. Black spots are normal signs of oxidation.

EXTENDING RIPENESS

An artichoke will keep refrigerated for up to two weeks. Sprinkle it with water—but do not wash it—before refrigerating the artichoke inside a resealable plastic bag.

RIPE. Ripe artichokes are 3" in diameter; harvest a ripe artichoke with a sharp knife.

OPTIMAL PLANTING AND HARVEST TIMES
Artichokes
Artichokes are perennial, and take as long as six months to mature.

BERRIES

BERRIES ARE DURABLE AND GENERALLY EASY to grow, although many plants have to grow into the second year before yielding a harvest. All berry plants produce abundant fruit.

Birds and wildlife prize berries, so it's wise to cover bird favorites like blueberries with protective netting. Pick berries in the morning, before the heat of the day.

RIPE. Ripe blueberries are a uniformly deep, dusty gray-blue. Taste blueberries to test ripeness—a ripe blueberry should be full-flavored and delicious.

BLUEBERRIES

The three main types of blueberry plants are: lowbush, half-high, and highbush. All three are harvested in the same way.

Color is key. Ripe blueberries are uniformly deep, dusty gray-blue. Allow the berries two days longer to mature after they turn the ripe color.

Pick them easy. Ripe blueberries almost fall off the bush and into your hand.

A taste test is a great way to determine ripeness. A ripe blueberry should be so plump and full that it almost bursts with juiciness on the tongue. It should be full-flavored and the flesh should be slightly firm but pleasingly tender—not mushy.

BLACKBERRIES

There are early, mid-, and late-season blackberry cultivars, so harvest time can vary. But the way they grow does not. Blackberry brambles grow new canes from the crown just below the surface of the soil, and from runners. Each cane lasts two years, producing leaves in the first year and fruit in the second.

OPTIMAL PLANTING AND HARVEST TIMES

Blueberries

Blueberry bushes can take three to four years or more to mature and produce fruit. Harvest time will depend on variety and climate, but the ripeness season is quite long, from April to late September.

Blackberries

Blackberry bushes take about 2 years to mature and produce fruit, but then can bear fruit for up to 15 years. Harvest time will depend on variety and climate. The prime season for ripeness is mid- to late summer.

Raspberries

Raspberry canes will produce good fruit beginning in the second year after planting. Ripe fruit is usually available four to six weeks after the start of the local growing season.

RIPE. Ripe blackberries (bottom right) are shiny and deep black with no areas of red.

If you have an "upright" blackberry variety, the fruit will develop at the tips of the canes. Trailing varieties will fruit out all along the length of the canes. Look for blooms on the plant, which signal that harvest is close. Tiny red "drupes" will form from the flowers, and ripe fruit follows within two to three weeks.

Color is the best measure of ripeness. The berries are shiny, deep black, with no areas of red. A ripe blackberry should taste subtly sweet; a red blackberry will be unpleasantly tart. Gently tug a ripe berry from the plant. If you're growing a thorny variety, wear gloves. Soft, shriveling, or moldy blackberries should be removed and discarded to prevent the spread of disease.

RASPBERRIES

Harvesting ripe raspberries starts with the type you're growing. Summer-bearing produces a single crop in summer; everbearing provides smaller harvests in summer and fall.

Pick raspberries on a sunny day, in the morning after the dew has dried. A ripe raspberry will come off the vine with a gentle tug. A raspberry wasn't ripe if the green cap is still attached after it's been picked.

Check the shape. The berry should be full-bodied and softened slightly. At the point where the fruit begins to wither, it is past prime.

Tasting is the best way to determine ripeness, because raspberries turn the mature color (some cultivars grow gold or black raspberries) before they're completely mature. Test one, and then pick all the berries that look like that one. A ripe raspberry will be juicy, with a delicate sweetness and tart undertones. Unripe raspberries will be overwhelmingly tart.

RIPE. Ripe raspberries come off the vine with a gentle tug.

IN THE MARKET

Berries are seasonal, so you'll want to get best value for your money.

Farmers' markets or roadside stands are more likely than supermarkets to offer truly fresh, ripe berries. They may also let you taste a berry to test ripeness. The best berry containers allow you to inspect the berries all over. Bigger capacity containers increase the possibility of berries being squished from the weight of the berries above.

Blackberries and raspberries should be deep, lustrous black, and rich, matte red. Blueberries should be uniformly blue, often with a light silvery "frost" on them (the berries' way of protecting themselves from sun exposure).

Avoid containers that include crushed berries or have juice stains, or that have the green caps still attached. Avoid wrinkled or shriveled blueberries and keep an eye out for any signs of mold.

EXTENDING RIPENESS

All berries prefer to be cool. Get them inside and out of the sun as soon as possible. Berries should be refrigerated unwashed until use, but the clock will be ticking. Their refrigerated shelf lives (at about 34°F) are:

Blueberries: up to 10 to 12 days
Raspberries: 2 to 3 days
Blackberries: up to 4 days

Berries freeze well and will keep for up to a year in the freezer. Lightly rinse with cool water and leave in a colander to dry. Once dry, spread the berries in a single layer on a small baking sheet and freeze the berries. Once frozen, transfer to a sealed container and store in the freezer until you're ready to use. Use frozen berries in baked recipes or for sauces or jams.

CHERRIES

ALTHOUGH SWEET CHERRY VARIETIES RIPEN before tart cherry cultivars do, they're harvested the same way and neither ripen off the tree. Picking the cherries at the right time is crucial, because sweetness and flavor increases radically in the last few days of the cherry's maturation. Sweet or tart, you'll judge ripeness by the same factors.

Color. Different varieties mature to different shades of red (or, with Rainier cherries, mottled red and yellow), but the mature color will always be saturated and lush.

Taste. Tasting doesn't work for tart cherries, but it's a good way to check sweet cherry ripeness. Taste one that looks ripe and it should be pleasingly sweet and juicy. Harvest all the cherries that look like that one.

Ripe tart cherries come off the stem with a gentle tug, while immature fruits won't budge. To pick sweet cherries, hold the stem at the tip that connects to the wood spur and gently twist it from the wood spur. Be careful not to damage the spur.

Pick sweet cherries with the stem attached to help them last longer and prevent bleeding. If you're canning or cooking the cherries, this isn't as important.

Handle harvested cherries carefully, and don't overfill picking containers, which can damage the fruit.

IN THE MARKET

You'll only find cherries in season. Look for firm, heavy cherries with bright, shiny skins and fresh, green stems. Stemless cherries are fine as long as the skin is still firm, bright, and there's no wrinkling around the shoulders.

Rainier cherries (below) are covered with red and yellow blotches, and are less firm when ripe. If they have a pinkish blush, the cherries have seen a lot of sun and will be very sweet. You may find some with brown flecks, which are not a problem.

OPTIMAL PLANTING AND HARVEST TIMES
Cherries
Cherries grow on trees that take five to seven years before they begin to produce fruit. Fruit appears in late spring to early summer.

RIPE. Ripe cherries are always saturated and lush in color.

EXTENDING RIPENESS

Cherries need to be cool as soon as possible after you pick or buy them. Keep them in the refrigerator in a loose or perforated bag for up to three days—the temperature should be around 35°F. Don't wash them first, but do give them a bath before eating.

If you're planning on cooking with the cherries, they can be frozen, either whole or pitted. Freeze them on a small baking sheet or tin, spread in a single layer and not touching. Once frozen, transfer to a tightly closed container. They'll keep for months in the freezer.

10 INCREDIBLY BEAUTIFUL RIPE
FRUITS AND VEGETABLES

For dynamic combinations of looks and flavor, turn to these edible garden eye-catchers.

1. Purple basil.

There are many deep purple cultivars, including purple Genovese varieties that can be used in the same way as green Genovese can. Purple basil is a stunning ornamental accent in a mixed bed or border.

2. Rainbow swiss chard.

'Bright Lights' is the most common variety, but all feature subtly crinkled leaves in shades of bronze and green, growing on brilliantly colored stems from fire-engine red, to sundrop yellow, to bright white.

3. Chocolate tomato.

This intriguing tomato is deep brown and red. There are grape, cherry, and heirloom chocolate varieties. Include one in an ornamental bed or mixed container garden.

4. Scarlet Runner bean.

A prolific producer, the plant grows long vines that can climb or trail. Cover a fence in Scarlet Runners and you'll enjoy a cascade of orange-red flowers with heart-shaped leaves providing the perfect backdrop.

5. Ornamental peppers.

These pepper plants are handsomely bushy, with conical, shiny, multicolored hot peppers. The edible peppers grow erect, and the plant is perfect as filler in a flower bed or purely as an ornamental.

6. Alpine strawberries.

These little gems are miniature strawberries, surrounded by serrated leaves on a bushy plant that rarely produces runners. Alpine strawberries are excellent for pathway borders, window boxes, and container and rock gardens.

7. Pineapple sage.

Pineapple sage offers a bushy full growth, gorgeous scarlet fall flowers that attract butterflies, and a scent and flavor that tastes of pineapple. The flowers, which are edible, have the same flavor.

8. Romanesco broccoli.

It's cultivated like any other broccoli, but Romanesco is marked with tight spiral buds and an incredibly stunning neon lime green color. Grow Romanesco broccoli anywhere you would site ornamental cabbage.

9. Silver thyme.

This bed or border filler has small leaves that—depending on the variety—are edged or stippled with white on a light-green background. From a distance, the foliage appears streaked with silver.

10. Redbor kale.

The deep magenta of the leaves and stems of redbor kale hold the same flavor and abundance of nutrients other kales do, but the frilled leaves and amazing color are perfect for a foliage bed.

FIGS

FIGS DON'T RIPEN OFF THE TREE. THEY SOFTEN, but don't improve in flavor or texture. Unripe figs taste terrible, so picking them at the height of maturity is the best way to enjoy figs. It's easy to tell when they're ripe and it's better to pick a fig slightly overripe than immature.

Color changes noticeably as the fruit matures. Most varieties start out light green and progress to dark brown, dark purple, or dark, dusky green. Figs should ripen a day or two more after turning their mature color.

Position on the branch is a telltale sign of fig ripeness. The necks holding the fig to the branch will begin to wilt, and the heavy fruit will go from perpendicular to hanging down.

Feel the ripening fig. Fully ripe figs are soft, almost exactly between firm and mushy.

To harvest figs, cut the stem with sharp scissors, leaving a short stub attached to the fig. The stub will delay spoilage. Handle figs carefully because they bruise easily. If you prefer, you can carefully pull the ripe fig off the tree. Simply hold it by its base and pull up and away from the tree. If a fig comes off with very little effort, it's truly ripe.

IN THE MARKET

Shop for figs that are all fully colored. Inspect them for any bruises and avoid bruised fruit. It's okay if the fruit is oozing clear nectar, but do not buy figs if they have any signs of milky-white latex on them. The figs should be soft but not mushy, and the stems should still be firmly attached. If the stems are loose, the fig is overripe and close to spoiling.

RIPE. Fully ripe figs are soft, almost exactly between firm and mushy.

OPTIMAL PLANTING AND HARVEST TIMES
Figs
Fig trees take two to five years or more to mature enough to produce fruit. There are normally two seasons for ripe fruit: June and August–September.

RIPE. These ripe figs are deep purple; upon ripening, the neck holding the fig will begin to wilt and the fruit will hang down.

EXTENDING RIPENESS

Ripe figs are best eaten or cooked as soon as possible. They'll keep refrigerated for up to three days. Don't store them near fresh vegetables, because they will age and spoil the vegetables very quickly.

Figs can also be dried in a dehydrator or outside on screens in direct sunlight. Freeze them for long-term storage by rinsing the figs and letting them dry totally. Spread them on a small baking sheet and freeze them, then transfer to a covered container and return to the freezer. They'll keep for years when frozen.

GRAPES

THERE ARE THREE BASIC TYPES OF GRAPES: American varieties—many of which are meant as "table" or eating grapes; European varieties, which are mostly used for winemaking; and Mediterranean varieties that are grown in warm areas of the country and include mostly wine grapes, with some table cultivars.

Determining ripeness and harvesting are the same for all varieties, and none ripen off the vine.

Color is the first sign of ripening. It's important that you know the correct mature color of the grapes you're growing. Any color should be rich and deep, saturating the grape, and covering it uniformly.

Feel the grapes. Unripe grapes are overly firm; mature grapes are plump and can be easily crushed.

Taste grapes. You need to know what the variety you're growing is supposed to taste like. Table grapes will have a familiar flavor. Wine grapes may be tart, even when ripe.

To harvest, hold the cluster securely and cut it from the vine with sharp garden shears. Handle it gently to ensure against bruising.

UNDER-RIPE. Unripe green grapes on the vine.

IN THE MARKET

You'll find green, red, and black table grapes in stores. Each has a slightly different flavor, and comes in seeded or seedless varieties.

Green grapes should be a rich yellow green, red grapes should be uniformly mid-range red, and black grapes should have a deep, lustrous hue.

The stems should be green and pliable. The grapes should all be firmly attached to the stems.

Avoid grapes that are moldy, shriveled, wrinkled, or otherwise marred. If the grape isn't shiny, it's well past ripe. A dusty-white coating, called *bloom*, is a natural, protective device that the grapes produce to avoid drying out.

OPTIMAL PLANTING AND HARVEST TIMES
Grapes
Most grapevines produce fruit in their third season. Prime harvest time will depend on variety and climate.

RIPE. Ripe Sauvignon Blanc wine grapes are a uniform green.

EXTENDING RIPENESS

Chill grapes as soon as possible. They can be kept in the refrigerator for five to seven days after you've removed any overripe, withered, or diseased grapes. Left on the cluster, bad grapes can quickly spoil the entire bunch.

You can store grapes for longer periods if you have a cellar or basement that maintains temperatures around 35°F, with high humidity. Store them long-term in crates lined with clean, dry straw, making sure that the clusters don't touch. In the right conditions, grapes will keep for up to six weeks.

GRAPEFRUIT

GRAPEFRUIT VARIETIES COME IN WHITE OR red (and shades of red). Regardless, the longer the fruit is on the tree, the sweeter it will become.

Color change is the first sign of ripening. The skin starts out green and slowly matures to pale yellow with hints of pink. It's best to wait until it is fully colored yellow with some pink blushing.

Taste testing is a reliable way to determine ripeness. Pick the grapefruit by holding it and gently twisting. A ripe fruit will come right off the tree. If the flavor is fresh and rich, pick the other grapefruits with the same coloring.

IN THE MARKET

Shop for grapefruits that have no green on the skin. The fruit should be firm and heavy for its size, with glossy skin.

EXTENDING RIPENESS

Store newly harvested or purchased grapefruits on the counter for up to seven days. The fruit may last up to three weeks in a low drawer in your refrigerator.

If you've purchased under-ripe grapefruit, you can hurry the ripening process along by putting the grapefruit in a paper bag with bananas, and closing the bag tightly.

∧ **UNDER-RIPE.** Not-yet-ripe grapefruit is dark green.

> **RIPE.** Ripe grapefruit is pale yellow with hints of pink; pick ripe grapefruit by holding the fruit and gently twisting.

OPTIMAL PLANTING AND HARVEST TIMES
Grapefruit
These trees take at least three years of maturing before bearing any fruit.

KOHLRABI

KOHLRABI IS A MEMBER OF THE CABBAGE family. There are both purple and green varieties, but color is not an indicator of ripeness. This is determined solely by the size of the bulb (actually the swollen part of the stem).

The bulb should be harvested at about 3 inches in diameter. Err on the side of smaller, rather than larger, bulbs, which can be fibrous with an unpleasant flavor.

To harvest the vegetable, cut the bulb from the root at soil level.

IN THE MARKET

Kohlrabi in stores is usually sold trimmed of its leaves. The bulb should be firm and solid, and the stems still erect. If the bulb has any soft spots or any of the stems are beginning to wilt, pass on the kohlrabi.

EXTENDING RIPENESS

Before cooking with or storing kohlrabi, trim the leaves. They should be used immediately, although they will last up to three days in the refrigerator. Keep kohlrabi refrigerated in the crisper drawer for up to three weeks. If you've peeled and cut the bulb, wrap it in plastic wrap and it should keep for up to two weeks.

RIPE. Ripe kohlrabi should be picked at about 3" in diameter; to harvest, cut the bulb from the root at soil level.

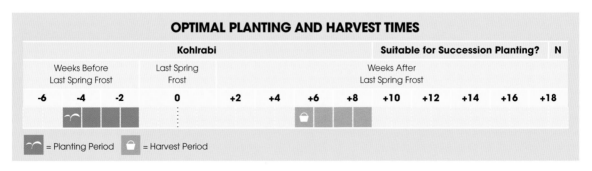

OPTIMAL PLANTING AND HARVEST TIMES													
Kohlrabi								**Suitable for Succession Planting?**					**N**
Weeks Before Last Spring Frost			Last Spring Frost		Weeks After Last Spring Frost								
-6	-4	-2	0	+2	+4	+6	+8	+10	+12	+14	+16	+18	

= Planting Period   = Harvest Period

LEMONS

ALTHOUGH THEY GROW ONLY IN THE MORE temperate areas of the country, lemon trees can be wonderful additions to a yard or garden. Smaller varieties can be grown in containers and moved to a sunroom or greenhouse over the winter.

Size matters. Lemons will ripen off the tree and are ready to be picked any time after they become mostly yellow. They should, however, be 2 to 3 inches in diameter. Ripe lemons have a glossy sheen.

Flavor remains the best sign of ripeness in lemons. Pick one you think is ripe and cut it open. Taste the flesh and juice; a full lemony flavor means it's ripe.

To harvest, hold the lemon firmly and twist gently until the stem breaks free from the tree. If you're concerned about damaging your tree, you can cut the stem with sharp garden shears or pruners. Harvest all the lemons on the tree if the temperature is going to drop below freezing.

IN THE MARKET

Look for full, well-proportioned fruit that is sunny yellow all over and heavy for its size. Small lemons can yield more juice on a per-lemon basis. Squeeze the lemon and it should give a little without being totally mushy.

Thicker lemon skins are better for zesting, while thinner-skinned lemons are usually juicier.

Avoid rock-hard lemons, and any with signs of mold, wrinkling, or soft spots.

If you're planning on using the lemon for zest, it's wise to buy organic, unwaxed fruit to ensure the skin hasn't absorbed any chemicals.

UNDER-RIPE. Don't pick lemons until they become mostly yellow.

OPTIMAL PLANTING AND HARVEST TIMES

Lemons

Citrus fruit trees can take two to six years to mature enough to bear fruit. Harvest season is normally the late fall to early spring seasons in southern climates.

RIPE. Ripe lemons will be at least 2 to 3" in diameter with glossy skin.

EXTENDING RIPENESS

Store slightly unripe lemons at room temperature until the lemon is completely yellow. Ripe lemons will keep on the kitchen counter—out of direct sun—for several days, and will last more than a week, refrigerated in a loose plastic bag. Bring lemons to room temperature if you're squeezing them for juice.

If you're unable to use all the lemons you harvest or buy, you can squeeze the lemon juice into ice cube trays and freeze the juice for use later. You can also freeze lemon zest in small, resealable plastic bags with the air squeezed out.

LIMES

A RIPE LIME CHANGES FROM DARK GREEN TO a lighter green; if a lime stays too long on the tree it turns yellow and becomes unpalatably bitter.

Picking fruit and cutting it open is the most reliable way to determine ripeness. If the lime is juicy and full flavored, you can start harvesting limes that look like the one you've picked.

Squeeze the lime. Ripe fruit should be slightly soft. The skin also becomes smooth when the fruit ripens.

Limes don't continue to ripen off the tree. Hold a ripe lime and give it a gentle twist and it should come off the tree.

IN THE MARKET

Look through the limes in the bin and pick those that are lighter green—they were picked riper. Any lime should have flawlessly smooth skin; buy the lime with the thinnest skin possible.

EXTENDING RIPENESS

Limes will last up to a week on the kitchen counter, if kept out of direct sunlight. They can last up to two weeks refrigerated in a loose plastic bag. Bring limes to room temperature before juicing them. You can also freeze lime juice in ice cube trays for later use.

RIPE. Picking ripe limes and cutting them open is the most reliable way to determine ripeness. Ripe limes should also be slightly soft.

OPTIMAL PLANTING AND HARVEST TIMES
Limes
Citrus fruit trees can take two to six years to mature enough to bear fruit. Harvest season is normally the late fall to early spring seasons in southern climates.

ORANGES

ORANGES TURN THEIR NAMESAKE COLOR WELL before they reach full maturity. Never pick green oranges, because oranges don't ripen off the tree.

The best way to test for ripeness is to peel and eat the orange.

Harvest an orange by holding the fruit firmly and giving it a gentle twist. It should come right off the tree. If for some reason picking the fruit this way causes the skin to rip, cut the stems with clean, sharp garden shears or pruners.

IN THE MARKET

An orange should feel heavy for its size. Thin-skinned oranges will be hard to peel but good for juicing. Large navel oranges are best for eating; find one with a pronounced crown—it will be easier to peel. Avoid bruised or dented fruit, or any that are overly soft or have even a hint of green on the skin.

EXTENDING RIPENESS

Oranges are best eaten fresh. They will keep up to a week stored on the kitchen counter, out of the sun. Refrigerate oranges in loose plastic bags, in a refrigerator drawer. Refrigerated oranges can last more than two weeks.

RIPE. The best way to test for ripeness with oranges is to peel and eat the orange. Harvest an orange by holding the fruit firmly and giving it a gentle twist.

OPTIMAL PLANTING AND HARVEST TIMES
Oranges
Citrus fruit trees can take two to six years to mature enough to bear fruit. Harvest season is normally the late fall to early spring seasons in southern climates.

PEACHES

PEACHES PICKED TOTALLY GREEN WON'T RIPEN. But harvested just a tad bit early, they can be ripened in a paper bag. The first peaches to ripen will be those near the top and outside of the tree.

Smell the fruit. A ripe peach gives off a sweet scent that you'll be able to smell if you're close enough to pick the peach.

Color is key. The ground color should be completely yellow on a ripe peach. The rest of the peach will be mottled with red blush. A ripe peach won't have any green undertones.

Taste a peach that looks ripe. It will be soft, flavorful, and juicy—but not mushy. There should be no crunch to the bite.

Be careful when harvesting to avoid bruising the fruit. Gently hold the peach and give it a slight twist, and it should come right off the tree.

IN THE MARKET

Buying local is a good idea. Peaches that don't have to travel as far are less likely to be bruised or overripe at the market.

Smell the peach. A ripe one will have a sweet aroma. There should be no green anywhere on the peach. It should also be soft, but without bruises or dents.

EXTENDING RIPENESS

Peaches are at their best when warm. Keep them on a counter until you're ready to eat them. They'll last up to three days at room temperature. Slightly green peaches can be ripened in a closed paper bag.

Refrigerate peaches in a loose plastic bag at around 34°F and they'll keep for up to two weeks.

RIPE. A ripe peach won't have any green undertones and will give off a sweet scent.

OPTIMAL PLANTING AND HARVEST TIMES

Peaches

Peach trees generally take two to four years before they are mature enough to bear fruit.
The ripe season ranges from mid-April in Florida to October in Idaho.

PEARS

MOST PEAR VARIETIES DO NOT RIPEN ON THE tree. The pears ripen from the inside out. By the time the skin is developed to its best texture, the core is essentially rotten.

Pick pears just slightly immature.

Ideally, you'll want to pick a pear when you are able to detach it by bending it horizontal from the position in which it hangs. Immature pears will be stubborn to remove. (However, Bosc varieties are always a bit difficult to separate from the branch's growing spur, whether ripe or not.)

If pears are dropping from your tree, you're already late in harvesting them. Consume the fallen pears first, cutting out any sections of rot, and quickly harvest the rest of the mature pears from the tree.

IN THE MARKET

Buying pears is a matter of determining when you'd like to use them. If you're ready to start eating them the moment you get home, do the thumb test on the pears. Hold the pear in the palm of your hand and gently dent the neck with your thumbnail. Because pears ripen from the inside out, the neck will indicate ripeness earlier than the body of the pear. If the neck indents with slight pressure, the pear is ready to eat. If it doesn't, the pear will take several days to ripen.

Avoid pears that are soft all over, bruised, or have obvious damage from handling. Handle pears gently at all times to avoid bruising them yourself.

EXTENDING RIPENESS

After harvest, pears ripen best if exposed to temperatures below freezing—usually about 30°F—for two days in the case of Bartlett pears, and several weeks for Anjou and Boscs (the high sugar content prevents the flesh from freezing). The cooling is part of the ripening process. They won't ripen the same by leaving them at room temperature.

Market pears are normally precooled before being shipped.

Speed ripening by putting pears in a paper bag with an apple or a banana for a day or two. You can also ripen pears on the counter, out of direct sunlight. Do not stack pears. Ripe pears will keep in the refrigerator for up to two weeks in a loose plastic bag.

> RIPE. Pears ripen from the inside out; they should be picked when they are still slightly immature.

OPTIMAL PLANTING AND HARVEST TIMES
Pears
Pear trees take four to six years to mature into fruit-bearing plants. Picking season varies by region, but begins in late summer and runs through fall.

PLUMS

REGARDLESS OF THE TYPE OR VARIETY YOU'RE growing, ripe plums all show the same indicators.

Feel the plum. Ripe plums begin to soften as they become juicier, while immature plums are completely firm.

Taste one that you think is ripe. It should be sweet and lusciously juicy.

Color is a reliable indicator of ripeness, but you have to know the exact color of your cultivar. Most ripe plums develop a "bloom," a white coating that protects the fruit from dehydration.

Plums are harvested by lightly but firmly holding the fruit and carefully twisting it off the stem.

IN THE MARKET

It's fine to buy plums slightly firmer than ripe, but avoid any that are wrinkled.

The fruit should seem heavy for its size, with a little give in the texture, especially at the end opposite the stem.

EXTENDING RIPENESS

Plums don't hold up well to aging. Eat them the day you buy or harvest them, or refrigerate. They'll last for up to a week in a plastic bag in a crisper drawer. Ripen slightly immature plums on the counter at room temperature for several days until they soften. Speed ripening by putting the plums in a paper bag with an apple or banana.

RIPE. Ripe plums begin to soften; if they're firm, they're immature. Another way to test ripeness is to taste one—it should be sweet and lusciously juicy.

OPTIMAL PLANTING AND HARVEST TIMES
Plums
Plum trees take three to six years to bear fruit. Prime picking season is between late spring and late summer in most regions.

RIPE EXOTIC FRUITS AT A GLANCE

If you're willing to experiment a little, you'll find these fruits present a completely new range of flavors, textures and—in many cases—scents. Many are also incredibly nutritious and some are even thought to have medicinal properties.

All of these are best eaten fresh. The flavor and other qualities rapidly deteriorate even if refrigerated. So enjoy these when you can, as soon as you can.

COCONUT

The fiber-covered husk contains a thick layer of white, sweet nut meat inside when fully ripe. The prized water inside is a product of immature coconuts.

Ripeness Indicator: Husk should be completely brown, and the coir (fiber) should be dry. Ripe coconuts will not have liquid inside. Listen for sloshing, and knock on the shell. It should have a hollow sound. Three "eyes" at the base should be soft and completely dry.

LYCHEE

Vibrant orange or crimson fruit with spiky skin and silky white flesh that tastes like a slightly acidic grape. The brown seed is poisonous.

Ripeness Indicator: Look for a full color and broad, plump shoulders. Small blemishes on the skin won't affect the quality of the flesh. Avoid dark colored or brown lychees, which are overripe.

MANGO

Large oval fruit with golden-yellow skin (often with blotches of red and green), dense golden, richly flavored flesh.

Ripeness Indicator: Stem end should smell sweet and like the fruit's flavor; texture should give slightly when squeezed. Color is not a reliable indicator of ripeness.

PAPAYA

Most common are Mexican and Hawaiian; Mexican papayas are larger. Both have green to yellow skin coloring, although Mexican may remain green even ripe.

Ripeness Indicator: The skin color should be largely yellow with a small percentage of green. The texture should feel slightly soft when gently squeezed—especially the end. Avoid any with large bruises, significant soft spots, or apparent shriveling.

STAR FRUIT

Also called Carambola, the fruit looks like a star when viewed from either end. When ripe, the flesh is succulent and tropical, like a plum with citrus tones.

Ripeness Indicator: Ripe star fruits have a light, almost translucent, yellow skin, waxy and vibrant. Ripe star fruit has a light, sweet, tropical aroma. May have brown lines along star "ridges."

DURIAN

With a spiky, thick, tough skin reminiscent of a hedgehog, and a strong smell that people either love or hate, the durian is the most unusual fruit on this list.

Ripeness Indicator: The unmistakable smell of a durian (it will either be enticingly sweet or more akin to "raw sewage") becomes strongest when ripe—the smell can be detected even through the skin. Listen to the fruit as you shake it; a ripe durian will clack like maracas.

KUMQUAT

A tiny member of the citrus family, the kumquat looks like a miniature, elongated orange. The fruit is eaten with the sweet peel intact, counteracting the tart flesh.

Ripeness Indicator: Color tells the story. Golden-orange, unblemished skin is the sign of ripeness.

PASSION FRUIT

This unusual fruit is shaped like a large egg and can be deep red-purple (the most common) or larger and yellow, and is prized for its sweet, tart pulp and seeds.

Ripeness Indicator: The skin color should be largely yellow with a small percentage of green. The texture should feel slightly soft when gently squeezed—especially the end. Avoid any with large bruises, significant soft spots, or apparent shriveling.

CACTUS PEARS

The name notwithstanding, the cactus pear is a berry, the fruit of the beavertail cactus. The two main types are a smaller, red-purple, and a larger, light green variety.

Ripeness Indicator: Look for small cactus pears that are richly colored. The fruit should be firm but not hard, and should give slightly as you press on it. The skin should be unblemished; avoid fruit with broken or scarred skin or any that is moldy on either end.

PERSIMMON

This golden-orange fruit comes in two basic types. The more common Hachiya is shaped like an acorn and ripens soft. The smaller, squatter Fuyu stays hard when ripe.

Ripeness Indicator: Ripe persimmons will be brilliantly orange, rounded and plump, glossy, and should have the leaf still attached. Hachiya should be slightly soft when ripe, while Fuyu persimmons will be firm. Avoid bruised fruit or any with yellow sections.

JACKFRUIT

The world's largest fruit, reaching 80 pounds in some cases. It's a lot of work to get to the edible flesh and seeds, but fans think the flavor (a cross between melon and mango) is well worth the effort.

Ripeness Indicator: The first sign of a ripe jackfruit is a nearly overpowering, heavy, fruity smell. The color of a ripe fruit will change from green to blotchy yellow. The fruit will change from firm to a texture that gives a little when you squeeze it.

GUAVA

These softball-sized fruits have a tender, understated flavor and mild sweetness that is coupled with a heavy, fruity scent. The skin is edible.

Ripeness Indicator: The skin color should be largely yellow with a small percentage of green. The fruit should feel slightly soft when gently squeezed. Avoid any with large bruises, as bruised guavas spoil quickly.

LONGAN

Similar to lychee, the brown, round longan is one of the most prized fruits in Thailand for its strawberry-like flavor and delicate texture.

Ripeness Indicator: When ripe, the longan's brown skin should be smooth and unbroken. There should be no mold and the fruit should not be mushy. If you peel or crack it open and the flesh is purple, brown, or any color but white, the fruit has spoiled.

POMEGRANATE

POMEGRANATES STOP RIPENING WHEN YOU pick them, but don't let them over-ripen, because the fruit is prone to splitting and will quickly lose its appealing tart flavor.

Color is a modest indicator of ripeness. There should be no green on a ripe pomegranate.

Feel the skin. It becomes leathery as the fruit matures. The ripe fruit changes shape from round to slightly hexagonal with distinct sections. Pomegranates get heavier as they mature.

Listen to the pomegranate. A ripe pomegranate will give off a metallic sound when you tap it.

Harvest ripe pomegranates by cutting their stems with sharp garden shears, leaving as little of the stem still attached to the tree as possible.

IN THE MARKET

Pomegranates are available seasonally. Buy fruit that is heavy for its size, with a deeply colored skin. Avoid pomegranates with a split skin or damage to the skin that exposes the interior.

EXTENDING RIPENESS

Refrigerate pomegranates in a crisper drawer; they will last more than a month at around 40°F. Freeze the arils for long-term storage by splitting the fruit and separating the arils from the stringy white pith. Spread the arils in a single layer on a small baking sheet and freeze them. Transfer to resealable plastic bags and they'll keep for months in the freezer.

RIPE. Ripe pomegranates show no green. Tap a pomegranate to determine its ripeness—it should give off a metallic sound.

OPTIMAL PLANTING AND HARVEST TIMES

Pomegranate

Pomegranate trees take 2½ to 3 years to bear fruit. Prime picking season is 5 to 7 months after bloom.

MARKET PRODUCE

THERE ARE SEVERAL FRUITS AND VEGETABLES THAT grow only in specific and limited climates in the United States, and are usually imported. That doesn't make us love bananas or avocados any less. As a matter of fact, the American public can't get enough of these delicious crops.

Fortunately, they travel well, and producers have developed many technologies and practices to help them reach your supermarket or local grocery store in good shape. You should still pay attention to what you're buying and make sure that the imported produce you purchase is as perfectly ripe as possible. This is even more important than it is with homegrown produce, because these imports generally cost a pretty penny.

AVOCADOS

THE HASS AVOCADO IS THE MOST FAMILIAR to grocery store shoppers, but it isn't the only avocado variety. You may come across others, but all share similar characteristics.

Hass. The Hass has thick, deeply pebbled skin that ripens to a purple-black and averages around 6 ounces. It has the typical creamy texture and luscious savory flavor.

Reed. Reeds average about 1 pound each and ripen to a deep green color. Although a good part of the fruit's weight lies in the oversized seed, there is still a lot of flesh to enjoy. Many growers consider this to be the best-tasting avocado of all.

Lamb Hass. Similar to the Hass, the Lamb Hass averages twice the size. The flavor and texture are the same, but the ripe color is a dark green.

Gwen. A rounder avocado, Gwen features flavor and smoothness similar to the Hass, but with a smaller yield of usable flesh because of the large seed.

Zutano. Zutano avocados are an elegant, pear-shaped variety with paler flesh and a noticeably lighter flavor than Hass. It has a thinner skin that stays mid-green as it ripens.

Regardless of variety, avocados don't normally ripen on the tree. Only after they are harvested do they change color and soften.

IN THE MARKET

You'll find avocados in various degrees of ripeness. Buy the perfectly ripe avocado to eat tonight, or pick up an almost-ripe one for guacamole this weekend. Any avocado that is mushy to the touch is far too overripe and is most likely rotten inside.

Color is a sign of ripeness. If you're not sure about the mature color of a given avocado in the produce department, check the firmness. All avocados become softer as they ripen.

RIPE. Hass avocados ripen to purple-black. The Hass has thick, deeply pebbled skin.

OPTIMAL PLANTING AND HARVEST TIMES

Avocados

Avocado trees must mature for five years or more before they will bear edible fruit. The avocado season is a long one in regions where they grow, extending from late March to mid-September.

RIPE. A Reed avocado ripens to a dark green color.

Feel the avocado but don't poke it. Hold it in your palm, with the fingers together and the thumb held loose, and very gently squeeze without using your fingertips. If it yields, it's ripe. If the avocado is greener and yields very little, it's about three to four days from ripe, and can sit on a counter until it is. If the avocado feels mushy, leave it in the bin.

EXTENDING RIPENESS

Ripe avocados can be refrigerated unpeeled for two days. If you're refrigerating a cut avocado, coat the cut surface with citrus juice and wrap tightly in plastic wrap. It will stay fresh for two to three days.

BANANAS

BANANAS CAN BE PICKED WHEN THEY ARE immature and transported while they ripen. This amazingly sweet fruit not only ages well, it is loaded with magnesium and potassium, and good amounts of fiber and vitamins.

IN THE MARKET

Bananas are all about the color. Green bananas are unripe (they are more starch than sugar and can be hard to digest), but will ripen on your counter. Size doesn't affect flavor or ripening.

Bananas that are a rich yellow are ready to eat and will keep for several days. Those with a scattering of brown spots are at the height of

RIPE. Ripe now, but these bananas will need to be eaten quickly before they overripen.

ripeness and ready to eat immediately. A brown banana is a supersweet baking ingredient, although these bananas are usually not eaten whole.

EXTENDING RIPENESS

It's easy to control the ripening process of the bananas you buy. Green bananas will usually ripen all by themselves on a counter within five to seven days. If you want to speed up the process, put green bananas in a paper bag with apples, close up the bag, and the bananas should be ripe and ready to eat in two to three days. Don't put green bananas in the refrigerator because the lower temperatures can short-circuit the ripening process and the banana may not continue ripening at all. However, you can refrigerate ripe bananas to slow their progress toward absolute browning. The peel will brown but the flesh will remain firm and yellow.

When you're ready to eat your ripe bananas, Chiquita recommends peeling them from the bottom (opposite the stem) up, so that you don't have to deal with the "strings."

Bananas freeze well for later use in smoothies and baking. Peel and slice bananas, and freeze the slices in plastic bags with the air squeezed out.

OPTIMAL PLANTING AND HARVEST TIMES
Bananas
These are exotic fruits imported from tropical climates, usually available year-round.

JICAMA

THIS LARGE ROUND ROOT IS COVERED BY a thick papery skin. The white flesh has the texture of an unripe pear and a savory, nutty flavor. The flesh is refreshingly crisp and crunchy. If you find this tuber locally, the quality will peak over the colder months, from later October to late April.

IN THE MARKET

The ideal jicama skin has a slightly silky appearance and the skin is perfectly smooth. Avoid any with damage to the skin that exposes the flesh, or with any major blemishes or noticeable soft spots.

Although you should buy jicama roots that seem relatively heavy for their size, don't buy large tubers. Medium or small tubers are less likely to be dried out when you get them home.

EXTENDING RIPENESS

A jicama with the skin on will keep in the refrigerator's crisper drawer for up to two weeks without appreciably drying out. If you're only using part of the tuber, wrap the remaining portion in plastic wrap and refrigerate for up to a week.

RIPE. Ripe jicama skin has a slightly silky appearance.

OPTIMAL PLANTING AND HARVEST TIMES

Jicama
These are exotic fruits imported from tropical climates, usually available year-round.

KIWI

KIWI (ACTUALLY CALLED "KIWIFRUIT") IS proof positive that big flavor can come in small packages. This strange little fruit is a vine berry. The fruit that seems to be a cross between the best of a melon and a strawberry is a produce aisle favorite.

exceedingly soft, and will be mushy and less flavorful when peeled.

Avoid kiwis with damaged skin, bruising, soft spots, or shriveling. If you can't find a ripe kiwi, buy the least-hard fruits you can find and ripen them at home.

IN THE MARKET

The skin of unripe and ripe fruit will look exactly the same.

Smell the fruit. Ripe kiwis have a faint, sweet fragrance.

Feel it, but don't squeeze it hard. A totally firm kiwi is unripe. The ripe texture is about the same as a ripe peach. Overripe kiwis are

EXTENDING RIPENESS

Kiwis will ripen on a counter at room temperature within three days. Speed the process up by putting them in a closed paper bag with an apple or banana.

Refrigerate them in a perforated loose plastic bag for up to a week.

RIPE. Squeeze a kiwi to test ripeness—totally firm is under-ripe.

OPTIMAL PLANTING AND HARVEST TIMES
Kiwi
These are exotic fruits imported from tropical climates, usually available year-round.

PINEAPPLE

FRESH PINEAPPLE IS ONE OF THE MOST exotic fruits we get to enjoy, thanks to modern food transportation. But to get the most out of what can be a relatively expensive produce purchase, it's essential to pick a ripe pineapple from those on display. Key to understanding pineapple ripeness is the fact that pineapples stop ripening when they are picked. It won't get riper over time, and it can be picked at absolute ripeness and still make the long-distance trip to your local market. A fully ripe pineapple will have abundant sugar that will make the fruit susceptible to bruising and rotting.

IN THE MARKET

Most shoppers have a false idea of the signs that determine a ripe pineapple. The pineapple should be plump, but not necessarily yellow. In fact, most of the pineapples you encounter will be more green than yellow, but they can still be totally ripe. If you have a choice, choose one that is yellow as far up from the bottom as possible (because the fruit ripens from the bottom up).

Pert green leaves in the crown are signs of freshness. (Don't fall for the old wives' tale that a loose center leaf signifies ripeness.) Size doesn't really matter in terms of ripeness or flavor, but you will wind up with more edible pineapple when you buy a larger fruit. A high crown (leaf cluster) is a good sign. Ideally, look for one that is at least 4 inches tall and as much as twice the height of the pineapple itself.

Avoid any pineapple with wrinkled skin or that is noticeably soft. And pass over any that smell of vinegar or have a chemical odor—signs that fermentation has begun.

EXTENDING RIPENESS

If you need to store a ripe pineapple, refrigerate it near the top of the compartment, ideally at a temperature around 45°F. When the pineapple is kept colder, the flesh can become waterlogged—more soupy than juicy. When kept at room temperature, a ripe pineapple can start to ferment.

RIPE. Pert green leaves in the crown of a pineapple are a sign of freshness.

OPTIMAL PLANTING AND HARVEST TIMES

Pineapple

These are exotic fruits imported from tropical climates, usually available year-round.

CHAPTER 3

MAKING THE MOST OF RIPENESS

AS IF IT WEREN'T ENOUGH OF A CHALLENGE TO GROW AND HARVEST perfectly ripe fruits and vegetables—and select and buy them—you also have to be sure you put your ripe produce to best use in the short time before things begin to decay and fade in quality. This is a bigger challenge than you might think. Experts at University of California at Davis estimate that 25 percent of all fruits and vegetables are not eaten because they are damaged or rot before they can be consumed.

> **25 percent of all fruits and vegetables are not eaten because they are damaged or rot before they can be consumed.**

Part of that is simply the reality of large-scale produce cultivation and transportation. A truck breaks down and peaches meant for the market are stuck on the road for an extra day. A freak rainstorm floods a lettuce field and makes picking impossible. But a lot of loss happens in the home, and much of that could be prevented.

It's a matter of knowing the fine points of how to handle the produce you grow or buy. It's just a fact that different fruits and vegetables are more or less sensitive to cold or heat. Some store well long-term, while some will barely make it a few days. Simple practices, such as when you wash a piece of fruit or a bunch of garden greens, can radically affect how long they will stay optimally ripe.

CARE AND HANDLING

Whether what you harvest is delicate or durable, you can never go wrong by treating produce with kid gloves from the moment you pick it to the moment you eat it. That means not dropping potatoes or apples into a bucket, but gently placing them instead. It means not stacking fresh-picked berries four layers high.

This is true, too, of the produce you buy. If you've ever watched as a grocery cashier tossed a peach into a bag of groceries, you've seen an assault on ripeness. If you're buying a lot of produce, or sensitive seasonal summer fruit, do your own bagging. Your family will thank you when they're eating wonderfully ripe tomatoes without squished sections, or apricots that haven't been bruised.

It's important to clean up dropped fruit around fruit trees to prevent the spread of disease and to keep on top of harvesting the ripe fruit.

I YAM WHAT I YAM

If you grew up referring to sweet potatoes as "yams," someone misled you. Yams are a very starchy tuber native to sub-Saharan Africa—and common in the West Indies and parts of Central and South America. Although they look something like the sweet potato when uncut, true yams are white inside and are very rarely grown in North America. They are sometimes sold in ethnic specialty stores, but are hard to find otherwise.

Keeping ripe produce safe also means—in most cases—keeping it away from diseased and damaged brothers and sisters. Diseased or rotting fruits and vegetables have the potential of passing those conditions on to healthy crops. Always police the ground and area around plants and trees, and quickly pickup and discard obviously defective fruit. Removing diseased fruits and vegetables from a plant not only limits the damage and spread of disease or insect infestation, it also helps spur future healthy growth.

Proper handling is only the first step in enjoying ripe produce. Just as important is how you store that produce once you get it into the kitchen.

STORING RIPE PRODUCE

Whether you're bringing a basket in from the garden or a brown paper bag in from the car, getting the most out of your ripe produce starts with where you put it when you unpack it.

Many fruits and vegetables begin changing the moment you pick them. Some start slowly, ripening even more, while others simply deteriorate in quality. Those reactions determine the best place to store any given fruit or vegetable in the short term.

Keep in mind that leftover sections of a fruit or vegetable are usually wrapped in plastic or put in a sealed container, then refrigerated. Of course, short-term storage isn't the only strategy for making best use of fruits and vegetables. There are many ways of preparing fruits and vegetables so that they can be used later and their flavors can still be enjoyed.

Juicing. Fruit—and some vegetables—that are on the cusp of overripe are often becoming soft and juicy. What better way to take advantage

There are two reasons for dropped fruit surrounding a tree or bush. It may be that you are late to the party and the fruit ripened so much that it simply fell off the plant. If this is the case and the dropped fruit appears healthy, start picking the ripe fruit from the tree or bush. Use dropped fruit first, because it will spoil more quickly than any you harvest from the tree. However, if dropped fruit is damaged or obviously diseased, pick it up and discard it so that the fruit doesn't spread the disease, and check the tree for other diseased fruit. If you find any, remove and discard it immediately.

If you have the space, time, and energy, putting up fruits and vegetables can be a fantastic way to extend ripeness throughout the year.

SHORT-TERM PRODUCE STORAGE AT A GLANCE

This chart will provide you with a quick-reference guide to where you should keep what you pick or buy so it stays at its best ripe quality. Particulars about how any fruit or vegetable should be stored will be found in the individual listings in Chapter 2. For instance, most leafy herbs will stay fresh longest placed stem-down in a glass of water with a plastic bag over it.

Crop	Fridge	Counter	Counter then Fridge	Pantry Notes
Apples	x			Separate from others
Artichoke	x			
Asparagus	x			Remove any ties
Avocados	x			
Bananas		x		Separate from others
Beans	x			
Berries	x			
Beets	x			
Bok choy	x			
Broccoli	x			
Brussels sprouts	x			
Cabbage	x			
Cauliflower	x			
Celery	x			
Cherries	x			
Citrus	x			
Corn	x			
Cucumber	x			
Eggplant	x			
Garlic			x	
Grapes	x			
Herbs	x			
Kohlrabi	x			
Leafy greens	x			
Melons			x	
Mushrooms	x			
Okra		x		
Onions (bulb)			x	
Pears		x		
Peaches		x		

Crop	Fridge	Counter	Counter then Fridge	Pantry Notes
Peppers	x			
Pomegranates		x		
Potatoes			x	No direct sun
Plums		x		
Squash (summer)	x			
Squash (winter)			x	No direct sun
Tomatoes		x		
Tropical		x		Pineapple, kiwi, etc.
Zucchini	x			

of that state than to reduce the fruit to its flavorful essence? You can juice many fruits by hand, although basic juicers are relatively inexpensive (and a wise purchase if you have a large citrus tree or two in your yard). The juice will keep for several weeks, if not months—a lot longer than ripe fruit or vegetables!

Preserving. Although it takes a fair amount of work, this is a tried-and-true method for keeping ripe fruit at the ready for whenever you want it. An incredible number of fruits and vegetables can be canned, pickled, fermented, or otherwise preserved at little cost and for great reward. A large tomato harvest can fill a pantry shelf and provide all the tomatoes you'll need to make it through winter. Pickle carrots and asparagus for a snack that is elegant enough to fill cocktail party platters, or ferment cabbage for the traditional Korean kimchi, a spicy favorite.

Drying. Many fruits and vegetables can be dried to preserve the ripe goodness. Drying, in most cases, concentrates flavor, maintains nutrients, and reduces the yield to a more manageable size for storage. It can also be free.

Sun-drying some of a bumper crop of tomatoes intensifies the flavor and creates a wonderful snack and ingredient that allows you to have the flavor of summer year-round.

KITCHEN WISDOM

Most herb plants are so prolific that they'll easily grow more leaves than you can use in a single season. Don't let the extra go to waste. You can make abundant amounts of herb vinegars, oils, and even butters, and the first two make excellent gifts. The trick is really to exploit the ripe flavors in their best medium, and make combinations that go well together.

Vinegar. The most obvious candidates for herbed vinegars are savory woody herbs, such as thyme, oregano, or rosemary. Simply cut a few sprigs for the vinegar, and they'll look as wonderful in a clear bottle as the vinegar tastes. But you can also make unusual vinegars with mint (perfect for dressing Mediterranean salads and fish dishes), dill, and tarragon. Judge the intensity of flavor you want and add the amount of herb you think will get you there to a clear stoppered bottle or lidded jar. Fill with warm vinegar and let steep for two weeks. Although white vinegar is the type most commonly used because it lets the herb flavor shine through, you can also experiment with red vinegar, champagne vinegar, or any other type. And, of course, you can combine herbs for totally unique flavors.

Herb oils and vinegars are wonderful ways to use up excess herbs, as well as to recycle interesting bottles. The condiments will keep for a long time and make excellent gifts.

Oils. Herb-infused oils are prepared differently than vinegars are, but the idea is the same: to capture the flavor of ripe herbs in a way that they can be stored. Generally, the base oil used is a light olive oil (not extra virgin), although you can try avocado oil, sesame oil, or any other cooking oil that might create interesting flavor combinations. For leafy herbs such as basil, tarragon, or flat-leaf parsley, use 2 cups of packed fresh leaves to every 1 cup of oil. For woody herbs such as rosemary, oregano, or sage, reverse the ratio. Combine the oil and herb in a blender and blend until smooth. Then heat the oil to a simmer for about a minute and strain the oil into a glass bottle or jar, through cheesecloth or a coffee filter. If the oil separates from a heavier liquid after a few hours, pour off the oil into a separate airtight bottle or jar.

Butter. Mixing up herb butter is a great way to use ripe herbs and create delectable butter for use on bread or steamed vegetables. Combine ¼ cup of your favorite herb (sage and rosemary are wonderful options) or a mixture of herbs, and 2 tablespoons of olive oil in a food processor. Blend on medium until combined. In a mixer, beat the oil with a pound of unsalted butter until smooth. Transfer the butter to a mold or to parchment paper and roll into a log. Chill for at least three hours before using.

You can dry fruits and vegetables on a screen outside. You can also use your oven to speed up the drying process. Of course, if you prefer, and you have a bumper crop in need of drying, you can buy a dehydrator to speed up and regulate the process.

These options all address possible short-term strategies, but there are longer-term alternatives that you can consider, depending on what you've harvested.

HOW TO SET UP A ROOT CELLAR

A root cellar is a traditional way to store fall crops for use over winter. Fortunately, you don't have to start excavating your backyard or move to the country to have one. That's because the cellar is more about climate than location. A corner of a basement can serve the purpose quite well. You just have to make sure the cellar will have certain essential features.

Humidity. Find the wettest corner of a basement or storage space, because most root cellar crops prefer humidity around 80 to 90 percent.

Ventilation. The space needs to have airflow. Usually a lower intake and higher exhaust vent are required for a functioning root cellar. The same is true if you're using a basement as one.

Shelving. The storage should be outfitted with shelves made from a material that will hold up to a wet environment and the stresses of supporting many pounds of produce. Ideally, they should be adjustable because the space will be cooler, lower to the floor, and different produce should be stored at different heights.

You can go further and create walls (with a high/low vent system) and a door for your root cellar space, which will allow you to more closely control the humidity and temperature. Once the space is together, you can begin storing your crops, following the recommendations in the individual listings in Chapter 2.

EXTENDING THE SEASON

Part of making the most of ripe produce is stretching the time when ripe is ripe. That's done by creating small or large artificial environments that create a longer growing season. These range from modest, such as simple row covers, to incredibly complex, as with a full-scale, fully equipped greenhouse.

Extending seasons won't work for every crop. For instance, it probably won't be possible to move a fruit tree into a greenhouse. But many of your plants can be tricked into believing fall extends into January, or that spring begins in February.

Row covers. When you just want to keep plants from freezing for a couple days to a

Row covers are ideal for raised beds too.

Cold frames are a simple way to extend the growing season.

Cold frames are angled to the south to take full advantage of the sun.

couple weeks, row covers can be the answer. They can also allow you to get a jump on plants that like to ripen in cool weather, but whose time to maturity will take them into summer weather. Row covers are inexpensive, widely available, and easy to use—just put up stakes or hoop supports and drape the cover over the row.

Hoop houses. Hoop houses are the midpoint between a row cover and a greenhouse. Enclosed, with a base and a skeleton of hoops, the hoop house allows you to get inside with your plants, and can

> A hoop house is a perfect choice for growing tall, bushy plants where a greenhouse is not practical.

accommodate taller or bushier plants. Larger hoop houses come with doors, making these the next best thing to a full-blown greenhouse.

Cold frames. Want to keep harvesting ripe salad greens into December? A cold frame may be just what you need. Cold frames are easy to construct in place over an existing garden bed. The structure is essentially a wedge-shaped outer frame angled to the south to take advantage of the sun, with a cover that can be as simple as plastic sheeting with a simple frame attached with hinges, or an operable recycled window. A cold frame is limited by the internal space.

Greenhouses. Greenhouses are the ultimate ripeness extender. A fully outfitted greenhouse will allow to grow crops year-round, and even a modest greenhouse will extend your growing to all but the coldest, darkest months. The downside is expense. But if you use a greenhouse every year and grow abundant crops that you might otherwise buy, you'll be amortizing the original cost over time and will probably realize value in the long run. Greenhouses are also a way to grow fruits that wouldn't necessarily ripen in your local climate. For instance, you can grow citrus trees in containers, and wheel them into the greenhouse when the weather turns freezing but the trees are still producing.

SUCCESSION PLANTING

A way to have the maximum amount of fresh, ripe produce from your garden and get the most return on your gardening investment is to plant crops in succession. This involves a good bit of planning, but can pay off in the long in ripe produce you won't need to buy from the supermarket (and that may well be better quality).

A permanent greenhouse is a solution for very serious gardeners.

You can plant successive crops of fast-growing fruits and vegetables such as radish and carrots, but the greatest potential in succession planting lies in cascading different crops, one right after the other. Do it right, and you'll have something ripe to enjoy from the garden from mid-spring through to early winter. Your garden will be busy for three seasons.

Use the charts in each profile to plan out your own succession gardening, using locally grown varieties of the plants that capture your interest. Do it right and you'll never have to ask, "Is it ripe?" Because at any one time, something surely will be.

INDEX

DEDICATED TO MEL BARTHOLOMEW (1931–2016)

THE BASIC PREMISE behind Square Foot Gardening seems simple today, but when Mel Bartholomew created it more than thirty years ago, it truly revolutionized the way home gardeners grow vegetables. Before Mel came along, most vegetable gardens looked pretty much the same: crops arranged in long rows separated by wide swaths of unplanted ground that consumed vast amounts of water and resources only to support the growth of weeds. His solution? Mel probably put it best:

> **"Build a box. Fill it with Mel's Mix. Add a grid . . . and start planting."**

Today, with more than 2.5 million copies of his Square Foot Gardening books in print, Mel has created a legacy dedicated to efficient, productive gardening and it has spread across the world. This new book, *Square Foot Gardening: Growing Perfect Vegetables*, is an extension of that legacy and is dedicated to his sincere efforts to bring fresh food into the lives of just about everyone on the planet. It is also a good example of how the SFG movement has grown and how its legions of dedicated followers are taking up the mantle to continue teaching the many benefits of planting food in Square Foot Garden boxes.

The Mel Bartholomew Foundation is a nonprofit 501(c)(3) corporation based in New York created to carry on Mel's work. And even though he was a born optimist and a tireless champion of his method, Mel would have been the first person to point out that there is plenty of work left to be done.

—*The Mel Bartholomew Foundation*